高等职业学校**美容化妆品专业**规划教材
美容化妆品行业职业培训教材

化妆品专业英语

Cosmetic English

李思彦 主编

化学工业出版社

·北京·

内 容 简 介

《化妆品专业英语》共设置了六个模块,选材来源于国外原版英文书刊、杂志、科技文献、国际机构网站信息等资料,收录了最新的行业信息、最实用的专业词汇和用语,保证教材的科学性、前沿性。内容包括化妆品市场、化妆品简介、化妆品原料、化妆品监管、化妆品安全、美容化妆品交际英语,各模块编写体例一致,由学习目标、课前导入、专业文献、重点词汇、实践活动、知识拓展(二维码)等部分组成,使教材既突出专业特色,又充分体现当代职业英语教育"教学练一体化"的理念,更有利于培养学生的技术应用能力及语言应用能力。

书后除总词汇表外,还以二维码形式介绍了英语构词法、化学元素和化合物的英文命名、化妆品颜色中英文对照。

本书既可供化妆品专业学生作为教材,又可供化妆品生产、管理、经营、销售及美容等从业人员阅读。

图书在版编目(CIP)数据

化妆品专业英语/李思彦主编. —北京:化学工业出版社,2020.10(2024.2重印)
ISBN 978-7-122-37593-3

Ⅰ.①化⋯　Ⅱ.①李⋯　Ⅲ.①化妆品-英语-教材　Ⅳ.①TQ658

中国版本图书馆CIP数据核字(2020)第158781号

责任编辑:张双进　刘心怡　　　　　　装帧设计:王晓宇
责任校对:边　涛

出版发行:化学工业出版社(北京市东城区青年湖南街13号　邮政编码100011)
印　　刷:北京云浩印刷有限责任公司
装　　订:三河市振勇印装有限公司
787mm×1092mm　1/16　印张19¼　字数475千字　2024年2月北京第1版第3次印刷

购书咨询:010-64518888　　　　　　　　售后服务:010-64518899
网　　址:http://www.cip.com.cn
凡购买本书,如有缺损质量问题,本社销售中心负责调换。

定　价:58.00元　　　　　　　　　　　　　　　　　　版权所有　违者必究

New Practical Cosmetic English

新编化妆品专业英语
编写人员名单

主　编　李思彦　广东食品药品职业学院
副主编　刘苏亭　潍坊职业学院
参　编　付尽国　广东科贸职业学院
　　　　张佳阳　河南应用技术职业学院
　　　　邹颖楠　广东食品药品职业学院
　　　　李晶晶　广东食品药品职业学院

前　言

随着近几年我国化妆品行业的高速发展，中国的化妆品企业正逐步实现与国际接轨，越来越多的企业对专业技术人员的外语水平提出了较高的要求。因此，化妆品从业人员的外语能力亟待提升。《化妆品专业英语》旨在提升化妆品从业人员、化妆品相关专业学生的专业英语水平，增强择业及就业竞争力。对于从事化妆品生产、经营、管理、销售、美容化妆等领域相关工作，本书有着重要的理论和实践指导意义。

本书通过参照化妆品技术专业、化妆品经营与管理专业的"化妆品专业英语"课程标准及相关课程的教学大纲，基于化妆品专业群的职业需求和典型工作任务，共设置了六个模块：化妆品市场、化妆品简介、化妆品原料、化妆品监管与法规、化妆品安全、美容化妆品交际英语。各模块编写体例一致，由学习目标、课前导入、专业文章、重点词汇、实践活动、知识拓展等部分组成，使教材既突出专业特色，又充分体现当代职业英语教育"教学练一体化"的理念，更有利于培养学生的技术应用能力及语言应用能力。本书具有以下几个特点。

（1）图文并茂，趣味性强。为了提高阅读的乐趣，本书在每课的课前导入、专业文章、实践活动、知识拓展等部分均搭配了丰富的相关图片，以达到浅显直观、深入浅出的效果。

（2）以任务为导向，互动性强。全书共设计了27节课，每课包含两到五个任务点，课前导入与各个任务点相关，通过图文并茂的问题讨论和相关词汇导入各任务点，使师生共同探讨每节课的学习内容。

（3）构思创新，实践性强。每节课的实践活动形式多样，设计了包括阅读理解、图文填空、中英翻译、词汇连线、小组讨论等题目，交际模块还设计了对话练习及听力练习，以提高本书实用性。

（4）选材新颖，专业性强。为了采录最地道的英文文献，本书的选材来源于国外原版英文书刊、杂志、科技文献、国际机构网站等，收录了最新的行业信息、最实用的专业词汇和用语，保证教材的科学性、前沿性。

（5）内容丰富，学习方式灵活。在每课的知识拓展部分，本书提供了大量相关知识的中文资料或英文短文，在全书附录部分提供了化妆品专业词汇的中英文对照，方便英语基础较好或有兴趣提高阅读能力及知识面的学生和读者自学。

（6）以职业技能为核心，情景教学。在美容化妆品交际模块，根据美容师、美体师、化妆师、美甲师等从业人员岗位需求，及美容前台、美容顾问、产品导购员等职业为线索，以美容院、化妆品门店工作过程为场景，培养专业人员与客户日常会话和专业交流的能力。

本书的编写队伍的作者均是来自全国职业院校的一线教师。本书由广东食品药品职业学院李思彦担任主编，主要编写分工如下：李思彦编写 Unit 1 和 Unit 6，广东食品药品职业学院李晶晶和邹颖楠合编 Unit 2，潍坊职业学院刘苏亭编写 Unit 3，广东科贸职业学院付尽国编写 Unit 4，河南应用技术职业学院张佳阳编写 Unit 5，李思彦和刘苏亭合编附录，李思彦负责统稿。

本书的编写得到了化妆品专业教师刘纲勇、杨梅老师的热心指导与帮助，也得到了安植集团筑梦美妆学院的支持，本书的出版也融入了化学工业出版社编辑的心血，在此一并表示衷心的感谢！

由于编写时间仓促及编者水平有限，教材中不妥之处在所难免，恳请广大读者批评指正。

编者

2020 年 5 月

CONTENTS

Unit One Cosmetic Market（化妆品市场）	**001**
Lesson 1 Cosmetic Market Overview（化妆品市场概况）	002
Task 1 China's Cosmetic Market（任务一 认识中国的化妆品市场）	003
Task 2 The History of Cosmetics（任务二 化妆品的发展历史）	010
Task 3 The Future Market Trends of Global Beauty Industry （任务三 全球美业未来的发展趋势）	014
Lesson 2 Brand Introduction of Cosmetics（化妆品品牌介绍）	023
Task 1 A Global Famous Beauty Brand—L'Oréal （任务一 认识一个世界知名的化妆品品牌——欧莱雅）	024
Task 2 HERBORIST: A Chinese Premium Brand of Cosmetic （任务二 认识一个中国高档化妆品品牌——佰草集）	032
Lesson 3 Cosmetic Marketing Strategy（化妆品营销策略）	036
Task 1 Cosmetic Online Marketing Strategy: Customization and Personalization（任务一 化妆品网络营销策略——定制化 与个性化）	037
Task 2 China's Cosmetics Top Brand Marketing Strategies to Unlock the Beauty Market（任务二 解读中国化妆品顶级品牌营销策略）	043
Unit Two Introduction of Cosmetic Products（化妆品简介）	**050**
Lesson 4 Skin Care Products（护肤品）	051
Task 1 Cleansers（任务一 认识洗面奶）	052
Task 2 Moisturizers（任务二 认识保湿霜）	053
Task 3 Body and Hand Creams and Lotions（任务三 认识霜和乳液）	054
Task 4 Toners（任务四 认识爽肤水）	054
Task 5 Eye Cream（任务五 认识眼霜）	056
Lesson 5 Color Cosmetics（彩妆产品）	061
Task 1 Facial Makeup（任务一 认识脸部化妆品）	062
Task 2 Eye Makeup（任务二 认识眼部化妆品）	065
Task 3 Lip Products（任务三 认识唇部化妆品）	068

Lesson 6　Hair Products（护发品）·· 071
　Task 1　Hair Care Products（任务一　认识护发产品）······························ 072
　Task 2　Hair Dye and Hair Colour Products（任务二　认识染发产品）········· 075
Lesson 7　Fragrance Product（香水产品）··· 077
　Task 1　Perfumes（任务一　认识香水）·· 078
　Task 2　Essential Oils（任务二　认识精油）·· 080
Lesson 8　Sunscreen（防晒霜）·· 083
　Task 1　Sunscreen: How to Help Protect Your Skin from the Sun
　　　　（任务一　了解防晒产品如何保护皮肤）································· 084
　Task 2　Understanding the Sunscreen Label（任务二　认识防晒霜标签）······ 087

Unit Three　Cosmetic Ingredients（化妆品原料）················ **091**

Lesson 9　Cleansing Ingredient（清洁类原料）··· 092
　Task 1　Cocamidopropyl Betaine（任务一　认识椰油酰胺丙基甜菜碱）······ 093
　Task 2　Other Cleansing Ingredients（任务二　认识其他常见清洁类原料）··· 095
Lesson 10　Skin Care Ingredient（护肤类原料）··· 101
　Task 1　Palmitates（任务一　认识棕榈酸酯类）····································· 102
　Task 2　Dimethicone（任务二　认识聚二甲基硅氧烷）···························· 105
　Task 3　Glycerin（任务三　认识甘油）·· 109
Lesson 11　Color Additives（化妆品颜色添加剂）·· 112
　Task 1　Color Additives Used in Cosmetics（任务一　化妆品中使用的
　　　　颜色添加剂）··· 113
　Task 2　Color Additive Categories（任务二　颜色添加剂目录）················ 118
Lesson 12　Auxiliary Materials（辅助性原料）·· 122
　Task 1　Parabens in Cosmetics（任务一　化妆品中的防腐剂尼泊金酯类）··· 122
　Task 2　Other Auxiliary Materials（任务二　认识其他常见辅助性原料）····· 128
Lesson 13　Plant Extracts（认识植物提取物）··· 134
　Task 1　Alpha Hydroxy Acids（任务一　认识果酸）································ 135
　Task 2　Other Plant Extracts（任务二　认识其他常见植物提取物）············ 138

Unit Four　Administration & Regulation（化妆品监管与法规）············ **144**

Lesson 14　U. S. Cosmetics Laws & Regulations（美国化妆品法规）················· 145
　Task 1　FDA's Authority over Cosmetics（任务一　美国FDA对化妆品的
　　　　监管）··· 146
　Task 2　FDA Recall Policy for Cosmetics（任务二　美国FDA对化妆品的
　　　　召回政策）·· 150

Task 3　FDA's Voluntary Cosmetic Registration Program（VCRP）
　　　　　（任务三　美国 FDA 的化妆品自愿注册计划）……………… 152
　　Task 4　Cosmetics Labeling Regulations（任务四　化妆品标签管理规定）…… 153
Lesson 15　EU Cosmetics Regulation（欧盟化妆品法规）…………………… 158
　　Task 1　Regulation（EC）No 1223/2009（任务一　欧盟化妆品管理法规 No 1223/2009）……………………………………………………… 158
　　Task 2　Colipa（European Cosmetics Trade Association）（任务二　欧洲化妆品行业协会）………………………………………………… 162
Lesson 16　China's Import and Trade Regulations（中国进出口化妆品管理）………………………………………………………… 167
　　Task 1　Hygiene Licence of Imported Cosmetics（任务一　进口化妆品卫生许可证）……………………………………………………… 167
　　Task 2　Hygiene License, Record-keeping Certificate & CIQ Labels for Imported Cosmetics in China（任务二　中国进口化妆品卫生许可证、备案证书及 CIQ 标签）……………………………………… 173

Unit Five　Cosmetic Safety（化妆品安全）……………………… **177**

Lesson 17　Cosmetic & Personal Care Product Safety（化妆品及个人护理产品的安全管理）……………………………………………… 178
　　Task 1　What the FDA Says about Cosmetic Safety（任务一　法律是怎么规定化妆品安全的）……………………………………… 178
　　Task 2　Safety and Technical Standards for Cosmetics（任务二　化妆品安全技术标准）……………………………………………… 180
Lesson 18　Cosmetic Ingredient Review（CIR）（化妆品成分评估）………… 182
　　Task 1　Introduction of Cosmetic Ingredient Review（CIR）（任务一　了解化妆品成分评估）……………………………………… 183
　　Task 2　Learn to Check CIR Findings（任务二　学习查找化妆品成分评估文件）……………………………………………………… 184
　　Task 3　Safety Assessment of Cosmetic Ingredients（任务三　化妆品成分的安全评估报告）……………………………………… 188
Lesson 19　FEMA GRAS Program（FEMA 的公认安全、无毒项目）………… 190
　　Task 1　The Flavor and Extract Manufacturers Association of the United States（任务一　认识美国香精和提取物制造商协会）……………… 190
　　Task 2　FEMA GRAS Program（任务二　美国香精和提取物制造商协会的公认安全无毒项目）……………………………………… 191

  Task 3 Learn to Check GRAS Notice Inventory（任务三 学习查找公认安全无毒清单）……………………………………………………………… 192

Lesson 20 Shelf Life and Expiration Dating of Cosmetics（化妆品的货架寿命和保质期）……………………………………………………………… 195
  Task 1 Factors Affect Shelf Life（任务一 影响货架寿命的因素）…………… 196
  Task 2 How to Keep Your Cosmetics Safe（任务二 如何保持化妆品安全）………………………………………………………………… 197

Lesson 21 How to Safely Use Cosmetics（如何安全使用化妆品）………… 200
  Task 1 Are Tattoo Inks Safe?（任务一 如何安全使用纹身墨水）…… 201
  Task 2 How to Safely Use Eye Cosmetics（任务二 如何安全使用眼妆产品）………………………………………………………………… 202
  Task 3 How to Safely Use Nail Care Products（任务三 如何安全使用美甲产品）………………………………………………………………… 204

Lesson 22 Science Behind Cosmetic Safety（化妆品安全背后的科学）……… 208
  Task 1 Factors Considered in Safety Evaluation（任务一 评估化妆品安全应考虑的因素）……………………………………………………… 209
  Task 2 Product Safety in the Marketplace（任务二 市场上产品的安全）…… 211
  Task 3 Cosmetic Product Safety Report（CPSR）（任务三 了解化妆品产品安全评估报告）………………………………………………………… 212

Unit Six Beauty & Cosmetic Communicative English（美容化妆品交际英语）……………………………………………………………… **215**

Lesson 23 Customer Reception（客户接待）……………………………………… 216
  Task 1 Greetings & Understanding Needs（任务一 问候客户及了解需求）………………………………………………………………… 216
  Task 2 Make, Change or Cancel Appointment（任务二 制定、更改或取消预约）………………………………………………………………… 219
  Task 3 Post-service（任务三 售后服务）…………………………………… 222

Lesson 24 Skin Care Service（皮肤护理服务）………………………………… 225
  Task 1 Understanding Your Skin（任务一 认识你的皮肤）…………… 226
  Task 2 Facial Care（任务二 脸部护理）…………………………………… 228
  Task 3 Introducing Service & Products（任务三 服务及产品推荐）………… 230
  Task 4 Massage & Body Care（任务四 按摩及身体护理）……………… 234

Lesson 25 Selling Cosmetics & Perfume（销售化妆品及香水）……………… 238
  Task 1 Selling Cosmetics（任务一 销售化妆品）………………………… 239
  Task 2 Selling Perfume（任务二 销售香水）……………………………… 243

Lesson 26 Make-up & Hairdressing（化妆及美发）…………………………… 248
  Task 1 Makeup Service（任务一 化妆服务）…………………………… 249

Task 2　Hair Styling（任务二　发型设计）……………………………………… 251
Lesson 27　Manicure Service（美甲服务）………………………………………… 259
Task 1　Service Process of Nail Care（任务一　美甲服务流程）……………… 261
Task 2　Nail Service Dialogues（任务二　美甲服务对话）…………………… 264

词汇表 ……………………………………………………………………… **268**

附录 ………………………………………………………………………… **296**

参考文献 …………………………………………………………………… **297**

Unit One
Cosmetic Market
化妆品市场

Learning Objectives 学习目标

In this unit you will be able to:
1. Have an overview of cosmetic market information including China's cosmetic market, cosmetics history, and the future market trends of the global beauty industry.
2. Get to know a world-famous brand and a domestic famous brand.
3. Understand common cosmetic marketing strategies such as online marketing and branding strategy.
4. Study more cosmetic brands and their marketing strategies.

Lesson 1 Cosmetic Market Overview
化妆品市场概况

Lead in 课前导入

Thinking and talking:
1. How many types of consumer markets do you know?
2. How many kinds of cosmetic products are there in the market?
3. Which cosmetic market do you think has the most potential and the fast-growing trend in China? Why?

Related words:

high-end market	高端市场	mid-end market	中端市场
low-end market	低端市场	domestic	国内
foreign	国外	skincare products	护肤产品
make-up products	化妆产品	natural cosmetics	天然化妆品
hair products	发用化妆品	anti-aging products	抗衰老产品

Task 1　China's Cosmetic Market

任务一　认识中国的化妆品市场

Ⅰ. Market Overview

China's cosmetics sector has been growing at a fast pace in recent years, in tandem with the rapid development of the economy. According to Euromonitor, retail sales of skincare products in China reached RMB212.2 billion in 2018, while sales of make-up products totalled RMB42.8 billion, representing year-on-year growth of 13.2% and 24.3% respectively. The table below (Table 1) shows how retail sales of cosmetic products by wholesale and retail enterprises above a designated scale have grown since 2013.

Table 1　The Retail Sales of Cosmetic Products from 2013 to 2018

Year	Retail sales/RMB billion
2013	162.5
2014	182.5
2015	204.9
2016	222.2
2017	251.4
2018	261.9

Source: National Bureau of Statistics of China.

The current structure of China's consumer market of cosmetic products is as follows.
Skincare products: the fastest-growing sector in the cosmetics market.
Shampoos and hair-care products: a market niche becoming saturated, seeing growth decelerating.
Make-up products: this market far from saturated, particularly for enhancement items, such as colour correcting (CC) and blemish balm (BB) cream. Sales of eye make-up products have recorded significant growth in recent years.
Products for children: sales of products designed for use by children continue to soar.
Sunscreen products: this help ensure sales not to slow down during traditionally quiet seasons.
Anti-aging products: cosmetic products that help consumers stay looking youthful and fight aging are increasingly popular.
Sports cosmetics: many consumers who love sports and fitness pursuits are keen to look good as well. They use sports cosmetics that help prevent moisture loss and are anti-odour, anti-sweat, anti-bacteria, packaged in compact, portable sizes.
Cosmeceuticals: consumers are increasingly aware of products that combine cosmetic and pharmaceutical features, such as spot lightening cream, acne treatment lotion and acne

ointment.

Green/natural cosmetics: these contain natural or nutritional ingredients, such as aloe and vitamins.

China's skincare-products market is becoming more high-end. Euromonitor figures show that though the retail sales value of high-end skincare products was still below that of mass-market alternatives in 2018, the market share of the former has been rising gradually, from 25.3% in 2013 to 32% in 2018. Consumers favour major international brand skincare products, and spending habits are switching from being price-focused to being driven by quality and brands.

China's domestic cosmetics brands are performing very well and now have a market share of about 56%, the main reason for this being their expansion into second-tier and third-tier markets. They have also been vigorously developing online sales and boosting advertising on new-media platforms (WeChat and Weibo) to raise brand recognition.

The men's cosmetics sector is growing strongly. According to estimates by Euromonitor, the male skincare products market expanded by 7.8% year-on-year in 2018, with men's skincare and make-up products becoming increasingly popular. When it comes to cosmetics, men are mainly concerned with cleansing and dealing with oily skin. While facial cleansers make up the lion's share of the male cosmetics market, demand for specialty products such as masks, sun-blocks and those with whitening and moisturising functions is also on the rise. This demonstrates that male consumers are beginning to pay more attention to skin conditions such as aging and coarseness.

Cosmeceuticals, especially China's herbal cosmetics, are opening up a new sector in the cosmetics market. Cosmeceuticals only make up about 20% of the market on the mainland at present, whereas in Europe, the US, Japan and Korea, cosmeceuticals command a 50%~60% share. China's cosmeceuticals market appears to have plenty of potential for growth. With young consumers beginning to concern themselves more about the ingredients and quality of the products they buy, the age at which they start to purchase cosmeceuticals is becoming increasingly lower. While cosmeceuticals have medical properties, they are classified as cosmetics since there is still no official definition for the term "cosmeceuticals" on the mainland. According to the Regulations on Cosmetics Hygiene Supervision, no medical jargon or claims of medical efficacy should be used in cosmetics items' packaging or instructions.

Consumers can be grouped into three major tiers—upper, middle and lower, based on their preference for brands, quality and price, along with their purchasing power. Buyers of imported brand products in the high-end market are mostly high-income earners in large and medium-sized cities. Most are young and middle-aged women who prefer famous cosmetics brands from Europe, the US and Japan.

People are increasingly aware of cosmetics safety issues. A series of problems arising from unsafe products have put consumers, manufacturers and regulatory authorities on alert. The introduction of hygienic and safety technology standards for cosmetics should help regulate the behaviour of cosmetics manufacturers and protect the rights of consumers.

All-natural DIY cosmetics have gained popularity in recent years. Consumers buy the ingredients themselves and create tailor-made cosmetics and skincare products with their own formulas. Today, DIY cosmetics have become increasingly commercialised, mass-produced and sold through e-commerce platforms such as Taobao. com and Tmall. com. Nevertheless, DIY cosmetics for online sales tend to have quality problems. They do not meet the requirements stipulated in the Regulations on Cosmetics Hygiene Supervision and have not applied for cosmetics production and sales permits.

China's imports of major cosmetic products in 2018 are summarized in Table 2.

Table 2 China's Imports of Major Cosmetic Products in 2018

HS Code	Description	2018 /US$ million	YoY change /%
33030000	Perfumes and toilet waters	404	57.2
33041000	Lip make-up preparations	647	47.1
33042000	Eye make-up preparations	209	59.3
33049100	Powders, whether compressed or not	222	59.5
33049900	Others(including preparations for the care of the skin, suntan preparations, etc.)	8,810	71.7
33051000	Shampoos	265	44.7
33052000	Preparations for permanent waving	4	28.9
3053000	Hair lacquers	11	115.1
33059000	Others	254	61.4
33072000	Personal deodorants and antiperspirants	12	92.0

Source: Global Trade Atlas.

II. Market Competition

According to statistics from the National Medical Products Administration (NMPA), which is now under the State Administration for Market Regulation (SAMR), there were 4,933 enterprises qualified to produce cosmetics in China at the end of June 2019. Domestic brands are mostly concentrated in the mid-end to low-end market segments, while joint ventures and enterprises with foreign investment dominate the high-end segment.

Because of the rapid development of domestic cosmetics companies, domestic brands' market share is gradually growing and creating increasing competition with their foreign counterparts. Domestic cosmetics brands have been active in image-building in recent years, for example, promoting marine-based skincare products, natural plant-based skincare products and modern Chinese herbal skincare products. Their reputation is on the rise both at home and abroad.

Domestic companies are paying increasing attention to product development and quality. Domestic make-up products are also becoming more diversified. Domestic companies have successfully opened up a cosmetics market with Chinese cultural characteristics, launching items like limited editions of Chinese-style vanity gift boxes. Domestic brands are actively applying traditional Chinese medicine concepts and natural extraction methods in their development of skincare products, such as the Tai Ji and Yu Wu Xing series from Herborist. In the past, domestic brands emphasised value for money and mostly targeted second-tier and third-tier markets. Today, some large domestic companies have started to develop high-quality products, aiming to meet the demands of increasingly discerning consumers in the domestic mid-end and high-end markets.

The cosmeceuticals market is dominated by foreign brands like VICHY, La Roche-Posay, Freeplus and Simple. Some domestic brands, including Tongrentang and Herborist, have also ventured into the cosmeceuticals market and are gradually achieving growing recognition from consumers.

Children's skincare products is a sector with huge potential and an increasing number of international childcare heavyweights are eyeing the China market. Frog Prince, Pigeon, Yumeijing, Giving and Johnson & Johnson are major players in the children's market. Competition is expected to become increasingly intense.

III. Sales Channels

The main sales channels for cosmetics on the mainland include integrated e-commerce platforms, wholesale markets, supermarkets and department stores, dedicated counters, specialty chain stores, drugstores, beauty parlours and direct selling. Integrated e-commerce platforms, department stores and specialty stores are now the top three sales channels.

The 'dedicated counter' is a major traditional sales channel for cosmetics, adopted by most world-renowned brands. Dedicated counters play a huge role in brands' image building. According to iiMedia Research, brand and word-of-mouth are the two factors that mainland consumers care about most when buying cosmetics. Top global brands such as Lancôme, Estée Lauder, Chanel and Dior dominate the sales of cosmetics through dedicated counters on the mainland. Only a few domestic brands, such as Herborist, are able to compete with these giants.

Some brands expand their business by opening specialty stores—mainly directly operated specialty store and franchise store formats. Many multi-national cosmetics giants prefer directly-operated specialty stores, in which they can display brand image better, ensure quality of service and enforce unified, stable pricing. Franchise chain stores, however, are generally regarded to be the most effective format, involving the least input and achieving the highest rate of success.

Direct selling is a way of trading cosmetic products through distributors' personal networks. Direct selling companies reward distributors depending on the quantity of goods sold through their networks. Avon was the first brand to launch a direct-selling pilot programme after the Regulations for the Administration of Direct Selling were introduced in 2005. Direct-selling licences were later granted to Amway, Perfect, Longrich and others.

Selling cosmetics through drugstores has become a major feature in China's cosmetics market. While the cosmeceuticals market is now dominated by foreign players, a number of

local pharmaceutical companies have begun to make use of this sales channel.
Cosmetic products are also distributed through traditional, pampering and therapeutic beauty parlours, large and medium-sized high-end beauty spas, franchise chain stores, and grooming and hairdressing parlours.
The retail concept of a 'cosmetics supermarket' or 'one-stop-shop' is becoming more popular, with the entry of players such as Watson's, Sephora of France and Sasa.
Many foreign brands have entered the mainland market by acquiring domestic brands and using their distribution networks. For example, MiniNurse and MG were acquired by L'Oréal, TJoy by Coty, and Dabao by Johnson & Johnson. Some foreign brands are establishing their presence in the mainland market through online shopping platforms.
Fairs held in China provide an ideal channel for industry players to gather the latest information and to meet dealers. Some of the cosmetics fairs scheduled to be held in China in 2019-2020 are listed in Table 3.

Table 3 Selected Cosmetics Fairs in China

Date	Exhibition	Location
5-7 September 2019	China International Beauty Expo (Guangzhou)	China Import & Export Fair Complex, Guangzhou
17-19 October 2019	Chengdu China Beauty Expo	Century City New International Convention and Exhibition Centre, Chengdu
30 October-1 November 2019	Shanghai International Beauty, Hairdressing & Cosmetics Expo	Shanghai Everbright Convention & Exhibition Centre
13-15 May 2020	China International Beauty Expo (Shanghai)	National Exhibition & Convention Centre, Shanghai
19-21 May 2020	China Beauty Expo (CBE Shanghai)	Shanghai New International Expo Centre

Key Words & Phrases 重点词汇

sector['sektə(r)] n. 部门；扇形，扇区；象限仪；函数尺 vt. 把……分成扇形
in tandem with 同；同……合作
retail['ri:teil] n. 零售 v. 零售，(以某价格)零售 adv. 以零售方式 adj. 零售的
wholesale['həʊlseil] n. 批发 adj. 批发的 adv. 大规模地；以批发方式 v. 批发
billion['biljən] num. 十亿
saturated['sætʃəreitid] adj. 饱和的；渗透的；深颜色的
decelerate[di:'seləreit] vt. 使减速 vi. 减速，降低速度
enhancement[in'hænsmənt] n. 增加；放大
significant[sig'nifikənt] adj. 有重大意义的；显著的；有某种意义的；别有含义的；意味深长的
keen[ki:n] adj. 渴望；热切；热衷于；热情的；热心的；喜爱；(对……)着迷，有兴趣
anti-['ænti] pref. 反；反对；对立；对立面；防；防止
market share n. 市场占有率；市场份额

vigorous['vigərəs]　*adj.* 充满活力的；果断的；精力充沛的；强壮的；强健的
expand[ik'spænd]　*v.* 扩大，增加，增强（尺码、数量或重要性）；扩展，发展（业务）；细谈；详述；详细阐明
demonstrate['demənstreit]　*v.* 证明；证实；论证；说明；表达；表露；表现；显露；示范；演示
cosmeceutical['kɔzmə'sju:tikəl]　*n.* 药用化妆品
Regulations on Cosmetics Hygiene Supervision　化妆品卫生监督条例
commercialise[kə'mə:ʃəlaiz]　*vt.* 商业化；使商业化
National Medical Products Administration(NMPA)　国家药品监督管理局
State Administration for Market Regulation(SAMR)　国家市场监督管理总局
joint venture　合资公司；合资企业
dominate['dɔmineit]　*v.* 支配；控制；左右；影响；在……中具有最重要（或明显）的特色；在…中拥有最重要的位置；俯视；高耸于
e-commerce[i:'kɒmərs]　*n.* 电子商务
department store　百货公司；百货商店
specialty store　专卖店
dedicated counter　专柜
word-of-mouth　*adj.* 口头的；口述的
beauty parlour　美容院
acquire[ə'kwaiər]　*v.* 获得，（通过努力、能力、行为表现）获得；购得；得到
fair[fer]　*adj.* 公平的；美丽的，白皙的；晴朗的　*adv.* 公平地；直接地；清楚地　*vi.* 转晴　*n.* 展览会；市集；美人

Practical Activities　实践活动

Part A　Reading comprehension.

1. According to the passage, which product is the fastest-growing sector in the cosmetics market? （　　）
   A. Hair-care products　　　　　　B. Body-care products
   C. Make-up products　　　　　　D. Skincare products

2. According to Table 1, what's the retail sales of cosmetic products by wholesale and retail enterprises in 2018? （　　）
   A. RMB 261.9 billion　　　　　　B. RMB 251.4 billion
   C. RMB 212.2 billion　　　　　　D. RMB 222.2 billion

3. In China's skincare-products market, how are consumer's spending habits switching into? （　　）
   A. price-focused　　　　　　　　B. quality and brand focused
   C. ingredient focused　　　　　　D. product safety focused

4. Which markets are China's domestic cosmetic brands expansion into? （　　）
   A. 1st and 2nd-tier markets　　　　B. 2nd and 3rd-tier markets
   C. 3rd and 4th-tier markets　　　　D. 4rd and 5th-tier markets

Lesson 1 Cosmetic Market Overview

5. Which product enjoys the most market share in the men's cosmetics market? (　)
A. facial cleansers　　B. masks　　　　C. sun-blocks　　D. moisturising products

6. Which of the following information is NOT true about most high-end market's consumers? (　)
A. high-income earners　　　　　　　B. young and middle-aged women
C. in large and medium-sized cities　　D. in small-sized cities

7. According to statistics, how many qualified enterprises can produce cosmetics in China at the end of June 2019? (　)
A. 4,899　　　　B. 4,933　　　　C. 4,399　　　　D. 4,833

8. How is the cosmeceuticals market mainly dominated? (　)
A. By domestic brands　　　　B. By foreign brands
C. By consumers　　　　　　　D. By manufacturers

9. According to the passage, which product's market competition is expected to become increasingly intense? (　)
A. men's cosmetics　　　　　　B. women's cosmetics
C. children's skincare products　D. cosmeceutical's market

10. Which of the following is NOT the top three sales channels for cosmetics in China? (　)
A. e-commerce platforms　　B. department stores
C. specialty stores　　　　　　D. drugstores

11. What are the factors that mainland consumers care about most when buying cosmetics? (　)
A. brand and product　　　　　　　　　　B. brand and word-of-mouth
C. product quality and word-of-mouth　　D. product safety and word-of-mouth

12. What are the TWO main store formats for opening specialty stores ? (　)
A. specialty store and department stores
B. specialty store and drugstores
C. department stores and franchise store
D. specialty store and franchise store

13. According to Table 3, how many cities are scheduled to hold cosmetics fairs in 2019~2020? (　)
A. Two cities　　B. Three cities　　C. Four cities　　D. Five cities

Part B Translation exercise.

1. Sports cosmetics: many consumers who love sports and fitness pursuits are keen to look good as well. They use sports cosmetics that help prevent moisture loss and are anti-odour, anti-sweat, anti-bacteria, packaged in compact, portable sizes.

2. While cosmeceuticals have medical properties, they are classified as cosmetics since there is still no official definition for the term "cosmeceuticals" on the mainland.

3. People are increasingly aware of cosmetics safety issues A series of problems arising from unsafe products have put consumers, manufacturers and regulatory authorities on alert.

4. Domestic cosmetics brands have been active in image-building in recent years-for example, promoting marine-based skincare products, natural plant-based skincare products and modern Chinese herbal skincare products.

5. Domestic companies have successfully opened up a cosmetics market with Chinese cultural characteristics, launching items like limited editions of Chinese-style vanity gift boxes.

6. The retail concept of a 'cosmetics supermarket' or 'one-stop shop' is becoming more popular, with the entry of players such as Watson's, Sephora of France and Sasa.

Part C   Matching words.

Task 2   The History of Cosmetics

任务二　化妆品的发展历史

Cosmetic products are today part of our regular culture and fashion, but that was not always the case. The first human made cosmetics appeared in early modern civilizations some 6 thousand years ago as the way to enhance the appearance and odor of the human body, but the difficult manufacturing processes, harmful ingredients and their connection to the high ruling classes created the aura of exclusivity around them. For the long periods of time, cosmetic products were frowned upon in Western history, and even actively forbidden to be used by many organizations. This "dark" period of cosmetic use finally ended during the end of the 19th and early 20th century, when great advancements in manufacturing, new entertainment industries and faster changes enabled the rise of famous cosmetic brands and their widespread use.

The first archeological evidence of cosmetics comes from the excavated tombs of Ancient Egypt pharaohs, but historians are convinced that first natural made cosmetics were used by our prehistoric ancestors much before the rise of modern civilizations. 6 thousand year old relics from Egypt tell us that their royalty and high class enjoyed several cosmetic products, such as face creams, perfumed oils, eyeliners, hair paints, castor oil, lipsticks, and lip gloss. As the centuries and millennia went, Egypt chemists found a way to simplify the manufacturing process of cosmetics, but that did not manage to remove the aura of their "exclusivity". One of the most important causes for that was their badly formed recipes, which often included poisonous ingredients that could cause serious illnesses. However, even with that, cosmetics remained an important part of Egyptian culture and especially their burial rituals. Among all cosmetic products, cedar oil was considered to be the most sacred one, because it was used in the process of mummification. That process used 7 types of oils, which were also the basis for the Egyptian ritual magic and medicinal remedies for various illnesses.

Even though Egyptian priest guarded their cosmetic recipes from the neighboring "primitive" civilizations, Mediterranean trade of the 1st millennia BC soon brought Egyptian cosmetic products to the shores of the newly formed Greek and Roman civilizations. There, high fashion was important and many wealthy people wore wigs, white face powder, and women used red lipsticks and red oils to make their palms "younger". In Rome, a woman was not considered beautiful if she did not use facial cosmetics. Lipsticks, skin creams made from beeswax, olive oils and rosewater, powders, hair colors and many other beauty treatments were widely used in the period of 100 BC and beyond. They even had a special type of female slaves whose only task was to help their masters to be more beautiful. Their names live with us even today—Cosmetae.

When Christianity rose, Christian women started to celebrate their religion with jewelry and cosmetics. Even the Old Testament mentioned two kings who painted their eyelids sometimes around 840BC. However, with the fall of the Roman Empire, Europe entered into dark ages where harsh living conditions, poverty, illnesses and constant wars prevented the spreading of expensive and extravagant fashion trends. This meant that almost all traces of cosmetic products disappeared from the European culture, not only because of its scarcity but also because the Christian church actively prevented it's spreading. Isolation of Europe finally came to the end in the 12th and 13th centuries when warriors returning from the crusades brought with them exotic cosmetic items from the Middle East where they never went

out of fashion. This new influx of riches and knowledge from the east soon kick-started European renaissance, which transformed Europe into an advanced civilization. Fortune started moving from the wealthy down to the middle classes, the industry was rising, sciences and arts received much needed funding, and trade routes started spreading new fashions much more quickly than before.

Even with all the advances of the Renaissance, cosmetics received little attention from the general population. Some used hair coloring, painted eggs on their faces to remove wrinkles and used similar "old age" removal techniques, but the widespread use of face and hand cosmetics never took hold outside aristocracy. The only really popular period of time when cosmetics was well received was during and shortly after the reign of English Queen Elizabeth Ⅰ (1558~1603). Her unique fashion style of stark white faces and brightly colored lips captured the attention of royalty and aristocrats across England and France, but that lasted only for a short time. Soon after that cosmetics (especially highly visible facial and nail paints) became common among low class women, such as prostitutes. Nothing changed much between the late 17th century and mid-19th century. Cosmetic products were uncommon among the majority of European civilization, in some cases receiving the status of banned and absolutely inappropriate merchandise. The only exceptions were medical cosmetic remedies that were used by everyone but the poorest in the 18th century.

The dawn of cosmetic use finally arrived in the second part of 19th century when the industrial revolution and great advances in chemistry (chemical fragrances) enabled much easier and varied production of various cosmetic products. With a much lower price and chemical ingredients that were much less dangerous for health, cosmetics started gaining serious foothold. Some of the most famous cosmetic products from that time were rogue red lipstick (it symbolized health and wealth), zinc facial powder (much safer than previous lead and copper based powders) and eye shadow and eye sparklers.

The turning point in the western fashion came in 1920s when mass marketed cosmetic products finally became financially viable. And where profit can be found, there is the will to market and sell it. Photography, the cult of film actors and big marketing campaigns soon brought the fall of traditional Victorian fashion, enabling women of all ages to start wearing cosmetic products in the public. Early decades of cosmetics popularity in the west brought us many inventive products, such as Lip Gloss by Max Factor, synthetic hair dye and sunscreen by L'Oréal, suntan and red nail polish by Coco Chanel, and others.

After the World War Ⅱ and its period of heavy material rationing, cosmetic industry experienced its second renaissance. Countless new fashion trends were adopted, mostly being popularized by various movie actresses and musicians. Today, cosmetics industry is a multi-billion dollar business that stretches across the entire world, always finding new ways and fashion trends that sustain and ensure its growth.

Key Words & Phrases   重点词汇

civilization[ˌsivələˈzeiʃn]   n. 文明；文化
odor[ˈəudə]   n. 气味；名声
aura[ˈɔːrə]   n. 光环；气氛；（中风等的）预兆；气味
exclusivity[ˌekskluːˈsivəti]   n. 排外性；独占权；特有性
frown[fraun]   vi. 皱眉；不同意   vt. 皱眉，蹙额   n. 皱眉，蹙额
archeological[ˌɑːkiəˈlɒdʒikəl]   adj. 考古学的
excavate[ˈekskəveit]   vt. 挖掘；开凿   vi. 发掘；细查
tomb[tuːm]   n. 坟墓；死亡   vt. 埋葬

Lesson 1 Cosmetic Market Overview

pharaoh['feroʊ] n. 法老；暴君
prehistoric[pri:hɪ'stɔ:rɪk] adj. 史前的；陈旧的
ancestor['ænsestər] n. 始祖，祖先；被继承人
relic['relɪk] n. 遗迹，遗物；废墟；纪念物
castor oil 蓖麻油
millennia[mɪ'leniə] n. 千年期；千周年纪念日
burial['beriəl] n. 埋葬；葬礼；弃绝 adj. 埋葬的
ritual['rɪtʃuəl] n. 仪式；惯例；礼制 adj. 仪式的；例行的；礼节性的
cedar oil 香柏油，雪松油，杉木油，红桧油
mummification[,mʌməfə'keʃən] n. [医]木乃伊化
primitive['prɪmətɪv] adj. 原始的，远古的；简单的；粗糙的 n. 原始人
Mediterranean[,medɪtə'reɪniən] n. 地中海 adj. 地中海的
Christianity[,krɪsti'ænəti] n. 基督教；基督教徒；基督教教义
harsh[hɑ:ʃ] adj. 严厉的；严酷的；刺耳的；粗糙的；刺目的；丑陋的
scarcity['skersəti] n. 不足；缺乏
warrior['wɔ:riər] n. 战士，勇士；鼓吹战争的人
crusade[kru:'seɪd] n. 改革运动；十字军东侵 vi. 加入十字军；从事改革运动
exotic[ɪg'zɒtɪk] adj. 异国的；外来的；异国情调的
influx['ɪnflʌks] n. 流入；汇集；河流的汇集处
Renaissance[ri'neɪsns] 文艺复兴
prostitute['prɒstɪtju:t] n. 卖淫者；娼妓；妓女；男妓
foothold['fʊthəʊld] n. 据点；立足处
synthetic[sɪn'θetɪk] adj. 综合的；合成的，人造的 n. 合成物

Practical Activities 实践活动

Part A Fill in the time-line of cosmetic history.

No.	Year	Important Event
1	_____ years ago,	the first cosmetics appeared in _____.
2	In _____,	cosmetics was brought to _____ where a woman was not considered beautiful if she did not use _____ cosmetics.
3	Around _____, when _____ rose,	even kings sometimes painted their _____.
4	With the fall of the _____,	almost all traces of cosmetic product disappeared.
5	In the 12th and _____ century,	warriors brought exotic cosmetic items from the _____ where never out of fashion.
6	(1558~1603)	The only time when cosmetics was popular was during the reign of _____.
7	Between _____ and _____,	cosmetic products were uncommon among the majority of European civilization.
8	The dawn of cosmetics arrived in _____,	thanks to _____ and great advances in chemistry.

续表

No.	Year	Important Event
9	In _____, the turning point,	when mass marketed cosmetic products became financially viable.
10	After _____,	cosmetic industry experienced its _____ renaissance.
11	Today,	cosmetics industry is a _____ business.

Part B Translation exercise.

1. The first human made cosmetics appeared in early modern civilizations some 6 thousand years ago as the way to enhance the appearance and odor of the human body, but the difficult manufacturing processes, harmful ingredients and their connection to the high ruling classes created the aura of exclusivity around them.

2. The first archeological evidence of cosmetics comes from the excavated tombs of Ancient Egypt pharaohs, but historians are convinced that first natural made cosmetics were used by our prehistoric ancestors much before rise of modern civilizations.

3. Even though Egyptian priest guarded their cosmetic recipes from the neighboring "primitive" civilizations, Mediterranean trade of the 1st millennia BC soon brought Egyptian cosmetic products to the shores of the newly formed Greek and Roman civilizations.

4. This new influx of riches and knowledge from the east soon kick-started European renaissance, which transformed the Europe into advanced civilization.

5. Some used hair coloring, painted eggs on their faces to remove wrinkles and used similar "old age" removal techniques, but widespread use of face and hand cosmetics never took hold outside aristocracy.

知识拓展

口红的前世今生

Task 3 The Future Market Trends of Global Beauty Industry

任务三 全球美业未来的发展趋势

In the year and decade ahead, beauty brands must go beyond product, and contribute positively to the world. Forecast top 5 largest beauty & personal care countries in 2020s as below.

The future of beauty: overview

As we enter 2020s, the start of a new decade, the beauty industry must prepare to take on

Source: Euromonitor International. bn: billion.

its biggest role yet. The expectations on brands across all sectors are growing greater, but within beauty, consumers have raised the bar for everything from efficacy to ethics, and in the year ahead, their demands will evolve even further.

Jessica Smith, the senior creative researcher at the Future Laboratory, explains: "In 2020s, the beauty industry will be defined by its contributions to society, whether through actions to help the environment or messages of empowerment."

Cosmetics Business forecasts 5 Global Beauty Trends as follows.

Ⅰ. Trend 1: The Conscious Beauty Diet

From meat to make-up, consumers are buying less. As awareness grows of the impact that the amount and types of purchases have on the planet, it is clear that conscious consumerism is here to stay. In the UK, 79% of women and 71% of men are reducing their consumption in general, according to a 2019 Walnut Omnibus survey of 2000 people.

In beauty, this movement has become especially pronounced. Trends such as 'slow beauty', 'minimalist beauty' and 'skip-care' all point the same way: consumers are not only being drawn to buying less, they are uncovering the beauty benefits of using fewer products on themselves. Mintel data reveals that more than a quarter (28%) of British women have reduced the number of their facial skin care products.

Mindful minimalism

Whether it's beauty dieting or beauty fasting, brands in 2020 will need to support consumer shifts towards a 'less-but-better' attitude.

"Brands need to evolve with this concept understanding that consumers are looking to purchase fewer items that still deliver all of the results they want," explains Sarirah Hamid, the founder of beauty trend insights company Pretty Analytics.

And Japanese brand Mirai Clinical has a less-is-more philosophy. Its hero product Skin Balancing Sugar Oil contains sugar oil (squalene) as its only ingredient, and the brand will shortly be discontinuing its other face products.

Another example is US brand Illuum, with its "you deserve less" philosophy—fewer products, fewer ingredients and less skin stress. The skin care brand offers only six products, many of which contain just two or three ingredients each, which are designed to equip the skin with the tools it needs to perform the job it was designed to do.

"Over-using skin care has been shown to weaken the skin barrier and make it more susceptible to things like acne, dryness, sun and pollution damage, and more," says Jessica Yarbrough.

Fundamentally, brands need to adhere to the idea of 'reduce, reuse, recycle' in order to support consumers as they buy more mindfully, advises Nick Vaus, a partner at branding design agency Free The Birds.

Multifunctional products, removing unnecessary packaging, refillability and bring-back loyalty-based recycling schemes are further ways for brands to help consumers tread more lightly on the planet, while making their lives simpler and easier, adds Vaus.

Ⅱ. Trend 2: Bioengineered Ingredients

Fear over chemicals and what they do to the skin, body and the environment has fuelled one of beauty's biggest influences over the past two decades—the natural movement. As consumers switched to 'green' or organic alternatives, natural and engineered were at opposite poles.

But the boundaries are now blurring due to two reasons: consumers are realising that natural does not necessarily mean better or more environmentally friendly, and new materials are emerging that straddle the line between natural and engineered.

Forecasters believe this development will further evolve in 2020 and beyond. "There is a growing realisation among beauty consumers that natural is not necessarily the gold standard in the beauty industry," says Emily Safian-Demers, a trends analyst at Wunderman Thompson Intelligence.

The EU and FDA still have no legal definition of natural for beauty products, which means that brands can greenwash or tout natural ingredients with little regulation or oversight.

This has resulted in consumers closely scrutinising ingredient lists and questioning products that claim to be 'green' or 'clean', while at the same time prioritising product performance.

The rise in beauty consumers looking for products, services and brands that can deliver targeted information about their individual skin condition and health illustrates this, and the developments in this area are plentiful.

With consumers becoming increasingly discerning, and turning a keen eye to ingredient quality and efficacy, Safian-Demers says that: "simply labeling a product as natural is no longer going to cut it."

Elevating natural beauty

At the same time, the idea that natural is synonymous with environmentally friendly is being dispelled, says Jessica Smith, "While plant-based products may be biodegradable, harvesting them is leading to widespread deforestation."

Wunderman Thompson Intelligence's Beauty Tech Futures report points out that garden-fresh ingredients and low-fi production are no longer the only preferences in the natural category, as bioengineered components start to be used in beauty products.

"Increasingly, biocompatible lab-grown alternatives are being aligned with organic and sustainable solutions for the natural beauty enthusiast," says the report.

And Icelandic brand Bioeffect, which was created when it bioengineered a plant-based replica of epidermal growth factor (EGF) in greenhouse-grown barley, has a range of award-winning anti-aging skin care products including its best selling EGF Serum, and a recently launched Imprinting Hydrogel Mask designed to enhance the effects of the serum.

Another interesting innovation this year was the launch of the first biodesigned human collagen ingredient created for skin care formulations. HumaColl21 from US-based Geltor is a commercial human Type 21 collagen selected for its maximum biocompatibility with human skin cells.

Geltor's co-founder and CEO, Alex Lorestani says "HumaColl21 opens the door to not only better results, but a better overall process that can safely be used across beauty, the food and beverage industry, and beyond."

A perfect blend?

"The idea that natural beauty and science cannot go hand in hand is being challenged by brands that are optimising naturally derived ingredients both for efficacy and for sustainability reasons," says Jessica Smith.

So how will this trend evolve further in 2020? Safian-Demers says: "We can expect to see more brands and products that integrate science and technology to offer unprecedented insight into consumers' natural skin health and unique biological make-up."

Ⅲ. Trend 3: Impactful Beauty

Zero-waste beauty was a key trend in 2019, but in 2020, brands will be expected to go beyond reducing environmental harm and contribute positively, with products that enable consumers to feel good about buying them. Zero impact is not enough, brands will need to add value through a driving purpose.

Emily Safian-Demers explains: "Modern consumers are increasingly conscious of the impact of their spending power."

More and more consumers are purchasing for ethics and values over brand heritage or prestige. In the age of the ethical consumer, brands are finding that status and branding no longer hold as much sway as purpose and principles.

The many brands who have made pledges to reduce environmental harm, either by rethinking their packaging or production processes, are to be applauded, and Safian-Demers notes that these pledges have, to an extent, sufficed to capture consumers' trust.

Positive packaging

Packaging that contributes positively to the environment is also being explored by brands.

BYBI's bio-plastic tubes are not only carbon-neutral but the sugarcane actually takes in more CO_2 than is produced when manufacturing the tubes.

And luxury skin care and fragrance brand Haeckels' 'biocontributing' mycelium and seed paper packaging can be planted in the garden to add nutrients as it biodegrades and brings new plant life when the seeds germinate.

Haeckels' founder Dom Bridges says: "If shopping as a concept is to continue, it must on all levels create at least no waste, but in order to create true sustainability, every product we make needs to contribute back to the ecosystem.

Consumption is vandalism, but if consumption contributed to our ecosystem and also aided planet health then the whole game would change."

IV. Trend 4: The Gen Z influence

From baby boomers to millennials, every generation has brought challenges and opportunities for brands, but with Generation Z(Gen Z)—those born between 1995 and 2010—things are different.

According to Business Insider, Gen Z will soon become the most pivotal generation to the future of retail. And it's not just the fact that one in three people on the planet is now a Gen Zer, nor it is because of their spending power, which is already estimated at $143 billion in the US alone.

It is the impact that their worldview, and their view of themselves, will have on brands that wish to stay relevant in the years ahead.

They will change not only how brands speak to and connect with consumers, they will require them to play a new kind of role in their lives. Gen Z are the first digital-native generation, the most diverse and tolerant generation, and are confident in their self-expression: the emphasis is overwhelmingly on "being yourself".

And this is where the beauty industry has not only a major opportunity, but a necessary one, for brands to survive in the future.

"Using the right language, and being inclusive—this is not going to go away," says Sarah Jindal, the senior global beauty and personal care analyst at Mintel.

"Brands need to create safe environments for people to express themselves and be themselves in the way that they want to. We're so used to brands having a message they share, but we are seeing new brands becoming more open and keeping things flexible for consumers."

Express yourself

Over the next ten years, consumers will expect brands to support them in their multiple physical and digital identities: they will not want beauty brands to tell them what they should look like, but rather invite them into a community with like-minded people, to spaces where they feel they can belong and find a sense of direction and purpose.

And Chanel's Coco Game Centers, which opened in Tokyo, Shanghai, Singapore, Seoul and

Hong Kong last year, provided a playful space for both gamers and beauty fans to experience customised arcade games based around the brand's latest collection of lip and nail colours.

"More brands are hosting events and creating immersive experiences that people can go to explore and discover and we are seeing a sense of community being created as it brings together people who enjoy the brand. The brand then becomes a social catalyst to bring like-minded people together," says Jindal.

Curious and boundary-pushing Gen Z consumers will seek out social interactions and experiences in a virtual realm too, says WGSN's Gen Z: Building New Beauty report.

"Brands should partner with avatar influencers and invest in creative AI programmes to resonate with shifting consumer mindsets," it says, noting that avatar brand marketing is becoming part of the mainstream. For example, Japanese brand SK-II's new ambassador is an AI avatar called YUMI.

V. Trend 5: Preparing for 2030

The next decade will be monumental for beauty and personal care, as the fourth industrial revolution mainstreams while consumer behaviour fluctuates widely and demographics become increasingly unreliable.

The way consumers choose, purchase and interact with products on a daily basis will significantly change due to advances in software, hardware, apps and augmented reality. In retail, VR and AR, paired with data and movement tracking, will be used to enhance consumer experience and product interaction.

Brands will use biometrics to assess consumers' preferences and provide valuable customisations, while neuromarketing will provide brands with more ways to measure potential success before a product or campaign hits the market, from detecting changes in emotional states to capturing the non-conscious aspects of a consumer's decision making.

Building trust

Trust will come under fire not only due to information shared, but also in response to the marketing tactics employed by brands. People will rely on their own instincts and be more informed than ever before when making product choices due to the research they carry out.

Brand transparency and avoidance of misunderstanding around every aspect of their businesses and products, together with a responsible eco-ethical mission, will become critical, and this will propel the 'clean beauty' trend to evolve even further.

Yet consumers who want both organic and sustainably produced products will embrace lab-grown ingredients and their comfort with biotechnology will increase. An early example is seed-breeding company Equinom, which uses DNA sequencing and custom algorithms to improve nutrition and create better-functioning seeds.

As a result of these developments, and many more, the brand-consumer relationship will change irrevocably by 2030 as the consumer dictates what they want from beauty companies, and how they want to receive it.

Key Words & Phrases 重点词汇

efficacy['efɪkəsi]　　n. 功效，效力
ethics['εθɪks]　　n. 伦理学；伦理观；道德标准
empowerment[ɪm'paʊərmənt]　　n. 许可，授权
conscious['kɑ:nʃəs]　　adj. 意识到的；故意的；神志清醒的
consumption[kən'sʌmpʃn]　　n. 消费；消耗
minimalism['mɪnɪməlɪzəm]　　n. 极简主义
susceptible[sə'septəbl]　　adj. 易受影响的；易感动的；容许……的　　n. 易得病的人
multifunctional[,mʌlti'fʌŋkʃənl]　　adj. 多功能的
bioengineer['baɪəʊ,endʒɪ'nɪə]　　n. 生物工程师
blur[blɜ:r]　　n. 模糊不清的事物；模糊的记忆；污迹　　v. 使……模糊不清，变模糊；使暗淡；玷污，沾上污迹
realisation[riləɪ'zeʃən]　　n. 实现，完成
tout[taʊt]　　vt. 兜售；招徕
scrutinise['skru:tɪnaɪz]　　vi. 作仔细检查；细致观察　　vt. 细看；仔细观察或检查；核对
prioritise[praɪ'ɔrəˌtaɪz]　　vt. 给予……优先权；按优先顺序处理　　vi. 把事情按优先顺序排好
discern[dɪ'sɜ:rn]　　vt. 觉察出；识别；了解；隐约看见　　vi. 辨别
elevate['elɪveɪt]　　vt. 提升，举起；振奋情绪等；提升……的职位
synonymous[sɪ'nɑ:nɪməs]　　adj. 同义的；同义词的
deforestation[di:,fɔ:rɪ'steɪʃn]　　n. 毁林，采伐森林，森林开发，烧林
epidermal growth factor(EGF)　　表皮生长因子
integrate['ɪntɪɡreɪt]　　vt. 使……完整；使……成为整体；求……的积分；表示……的总和　　vi. 求积分；取消隔离；成为一体　　adj. 整合的；完全的　　n. 一体化；集成体
pledge[pledʒ]　　n. 保证，誓言；抵押；抵押品，典当物　　vt. 保证，许诺；用……抵押；举杯祝……健康
vandalism['vændəlɪzəm]　　n. 故意破坏他人（或公共）财物罪；恣意破坏他人（或公共）财产行为
millennial[mɪ'leniəl]　　adj. 一千年的；千禧年的
immersive[ɪ'mɜrsɪv]　　adj. 拟真的；沉浸式的；沉浸感的；增加沉浸感的
avatar['ævətɑːr]　　n. （印度教，佛教）神的化身；（某种思想或品质）化身；（网络）头像，替身
eco-ethical　　生态道德

Practical Activities 实践活动

Part A　Reading comprehension.

1. According to the passage, which country rank No. 2 among the forecast largest beauty & personal care countries in 2020s? (　　)
A. US　　　　　　B. Japan　　　　　　C. China　　　　　　D. Brazil

Lesson 1 Cosmetic Market Overview

2. According to Trend 1 The conscious beauty diet, the "you deserve less" philosophy from US brand Illuum means (   )?
A. more products, fewer ingredients and less skin stress
B. fewer products, more ingredients and less skin stress
C. fewer products, fewer ingredients and more skin stress
D. fewer products, fewer ingredients and less skin stress

3. According to Trend 1 The conscious beauty diet, in order to support consumers as they buy more mindfully, what should brands do? (   )
A. reduce   B. reuse   C. recycle   D. All of the above

4. According to Trend 2 Bioengineered ingredients, in order to offer consumers' natural skin health and unique biological make-up, what would more brands and products integrate? (   )
A. science   B. technology
C. Innovation   D. science and technology

5. What are those people born between 1995 and 2010, and will become the most pivotal generation to the future of retail, called? (   )
A. Generation X   B. Generation Y   C. Generation Z   D. Generation A

6. According to Trend 5, what will be used to enhance consumer experience and product interaction in retail? (   )
A. VR   B. AR
C. data and movement tracking   D. All of the above

7. According to Trend 5, why the brand-consumer relationship will change irrevocably by 2030? (   )
A. Because the consumer dictates what they want and how they want to receive it.
B. Because the beauty companies dictate what they want and how they want to receive it.
C. Because the products dictate what they want and how they want to receive it.
D. Because the brands dictate what they want and how they want to receive it.

Part B Translation exercise.

1. Trends such as 'slow beauty', 'minimalist beauty' and 'skip-care' all point the same way: consumers are not only being drawn to buying less, they are uncovering the beauty benefits of using fewer products on themselves.

2. "Over-using skin care has been shown to weaken the skin barrier and make it more susceptible to things like acne, dryness, sun and pollution damage, and more," says founder Jessica Yarbrough.

3. Multifunctional products, removing unnecessary packaging, refillability and bring-back loyalty-based recycling schemes are further ways for brands to help consumers tread more lightly on the planet, while making their lives simpler and easier.

4. With consumers becoming increasingly discerning, and turning a keen eye to ingredient quality and efficacy, Safian-Demers says that: "simply labeling a product as natural is no

longer going to cut it."

5. We can expect to see more brands and products that integrate science and technology to offer unprecedented insight into consumers' natural skin health and unique biological make-up."

6. Gen Z are the first digital-native generation, the most diverse and tolerant generation, and are confident in their self-expression: the emphasis is overwhelmingly on "being yourself".

Lesson 2 Brand Introduction of Cosmetics
化妆品品牌介绍

Lead in 课前导入

Thinking and talking:

1. How many world-famous cosmetic brands do you know?
2. How about Chinese cosmetic brands?
3. Try to recognize the brand logos as follows.
4. Which brands do you like best? Try to explain your reasons.

Task 1　A Global Famous Beauty Brand—L'Oréal

任务一　认识一个世界知名的化妆品品牌——欧莱雅

Ⅰ. Who We Are

For more than a century, we have devoted our energy and our competencies solely to one business: beauty. We have chosen to offer our expertise in the service of women and men worldwide, meeting the infinite diversity of their beauty desires. We are committed to fulfilling this mission ethically and responsibly.

Our Ambition

Our ambition for the coming years is to win over another one billion consumers around the world by creating the cosmetic products that meet the infinite diversity of their beauty needs and desires.

1. Beauty for all, beauty for each individual

At L'Oréal, we are convinced that there is no single and unique model of beauty, but an infinite diversity of forms of beauty, linked to periods, cultures, history and personalities... To draw ever greater numbers of women and men to use our products means reaching out to extremely diverse populations with a vision of universalising beauty. In our view, to universalise does not mean to impose uniformity, but on the contrary, to be inspired by diversity to innovate.

2. Observe local beauty customs

At the heart of this project, our Research and Innovation reinvents itself to create cosmetics products adapted to the immense diversity of the world. In each region of the globe, we have set up research platforms, true centres of expertise, designed with tailor-made beauty in mind. These research poles invent new products that can become worldwide successes. This is a real turning point in the way we think about innovation.

3. Facilitating access to cosmetics products

In a market undergoing substantial transformations, L'Oréal takes steps forward each year to make the best in beauty available to everyone. With a portfolio of 32 international brands and an organisational structure based on distribution channels, we have the ambition to meet the needs of every consumer according to his or her habits and lifestyle. In its own way, L'Oréal is thus pushing back boundaries and taking up the challenge of increasingly accessible innovation.

4. To accelerate the regionalisation of our expertise

To win over another one billion consumers around the world is an ambitious project that motivates all our teams. An economic but also human adventure, which requires the rapid deployment of our forces and an accelerated transformation of the company in every field, including research, manufacturing, marketing, sales, human relations and administrative teams...

This major project is also an opportunity for innovation and progress to build the L'Oréal of tomorrow.

Our Mission

1. Beauty is a language.

For more than a century, we have devoted our energy and our competencies solely to one business: beauty. It is a business rich in meaning, as it enables all individuals to express their personalities, gain self-confidence and open up to others.

2. Beauty is universal.

L'Oréal has set itself the mission of offering all women and men worldwide the best of cosmetics innovation in terms of quality, efficacy and safety. By meeting the infinite diversity of beauty needs and desires all over the world.

3. Beauty is a science.

Since its creation by a researcher, the group has been pushing back the frontiers of knowledge. Its unique research arm enables it to continually explore new territories and invent the products of the future, while drawing inspiration from beauty rituals the world over.

4. Beauty is a commitment.

Providing access to products that enhance well-being, mobilizing its innovative strength to preserve the beauty of the planet and supporting local communities. These are exacting challenges, which are a source of inspiration and creativity for L'Oréal.

5. L'Oréal, offering beauty for all.

By drawing on the diversity of its teams, and the richness and the complementarity of its brand portfolio, L'Oréal has made the universalization of beauty its project for the years to come.

Our Values and Ethical Principles

The founding values and ethical principles of the group are expressed in the daily operations of all our teams around the world.

Our Values

Our values are embedded in L'Oréal's genetic code. They continue to this day to express themselves in the daily actions of all our teams across the globe. Here is a close-up on the group's six founding values.

1. Passion

If for over a century L'Oréal has been devoted to just one business—beauty—it is above all because of our passion for it. Passion for what cosmetics can bring to women and men: well-being, self-confidence, an openness towards others. Passion also for a business is intrinsically linked to humanity and culture. Because creating beauty products means seeking to understand others, knowing how to listen to them, apprehending their traditions, anticipating their needs… Without this passion, the L'Oréal adventure would never have been possible.

2. Innovation

Innovation is also one of our founding values. We always have in mind the fact that our company was founded by a scientist. Innovation is essential because beauty is an endless quest that constantly requires a higher level of performance. At L'Oréal, it is vital. Always wanting to push back the limits of knowledge means discovering new ways to create products that are truly different and surprising. To always stay one step ahead.

3. Entrepreneurial spirit

Because there is no innovation without daring, without taking initiatives, L'Oréal has always given priority to the individual rather than to organisations. Entrepreneurial spirit, a synonym for autonomy, challenge, and adventure, has always been encouraged and embodied in a specific management style. Today it is still the driving force behind a group built above all on a belief of the importance of each individual and their talents.

4. Open-mindedness

Another value that has been guiding us since the group's foundation over a century ago is open-mindedness. Listening to consumers and understanding their culture, being open to others and benefiting from their differences are absolute priorities in order to respond to the infinite diversity of beauty aspirations around the world. They are inseparable from our business and our mission.

5. Quest for excellence

These four values are inextricably related to our fifth value, the quest for excellence. A value that permeates every aspect of our business, in every country and that is expressed in a state of mind and a constant pursuit of perfection. We all share this desire to surpass ourselves to be able to provide the best for our consumers.

6. Responsibility

Finally, whether the group is innovating or showing its entrepreneurial spirit, it has always done so with a sense of Responsibility. L'Oréal's first invention, the "safe hair dye" was already an expression of this fundamental concern for effective, safe and innocuous products. But our sense of responsibility goes far beyond that. As a world leader in beauty, we have, more than others, the duty to preserve the beauty of the planet and to contribute to the well-being of our employees and of the communities in which we are present.

Our Ethical Principles

Our principles are **Integrity, Respect, Courage** and **Transparency.**

Our Ethical Principles shape our culture, underpin our reputation, and must be known and recognised by all L'Oréal employees.

Integrity because acting with integrity is vital to building and maintaining trust and good relationships.

Respect because what we do has an impact on many people's lives.

Courage because ethical questions are rarely easy but must be addressed.

Transparency because we must always be truthful, sincere and be able to justify our actions and decisions.

II. History

For more than a century, L'Oréal has been involved in the adventure of beauty. The small company founded by Eugène Schueller in 1909 has become the number one cosmetic group in the world. Major launches, acquisitions, opening new subsidiaries... relive the highlights of the L'Oréal adventure.

1909-1956 : THE FIRST STEPS, CONSTRUCTING A MODEL

In 1909, Eugène Schueller, a young chemist with an entrepreneurial spirit, founded the company that was to become the L'Oréal group. It all began with one of the first hair dyes that he formulated, manufactured and sold to Parisian hairdressers. With this, the founder of the group forged the first link in what is still the DNA of L'Oréal: research and innovation in the service of beauty.

1957-1983 : "ON THE ROAD TO THE GRAND L'ORÉAL"

These are the formative years of "Le Grand L'Oréal".

At the instigation of Chairman François Dalle, the group starts to expand internationally. Acquisitions of strategic brands mark the beginning of a period of spectacular growth for the company. Emblematic products come into being.

The company motto is "Savoir saisir ce qui commence" (seize new opportunities).

1984-2000 : BECOME NUMBER ONE IN THE BEAUTY INDUSTRY

These twelve years are marked by a great period of growth for L'Oréal, mainly driven by the significant investments made by the group in the field of research.

Alongside these efforts are strategic product launches that not only make history, but also succeed in strengthening the group's brand image.

In 1988, François Dalle's successor, the research and development pioneer Charles Zviak, hands over the reins of the company to Lindsay Owen-Jones, a truly outstanding director. Under his management, the group would completely change in scope to become the world leader in cosmetics through the worldwide presence of its brands and strategic acquisitions.

2001-PRESENT DAY : DIVERSITY OF BEAUTY WORLDWIDE

There is no single type of beauty; it is a multiple-faceted quality framed by different ethnic origins, aspirations, and expectations that reflect the world's intrinsic diversity. With a portfolio of powerful, international brands, L'Oréal enters the 21st century by embracing diversity in its global growth agenda. Headed since 2006 by the chairman Lindsay Owen-Jones, and the chief executive officer Jean-Paul Agon, and then by Jean-Paul Agon who was appointed chairman and CEO in 2011, the group continues to make new acquisitions to cover the world's varied cosmetic needs, and to undertake new socially responsible initiatives in the interests of sustainable development for all.

III. What's in Our Products?

We know you care deeply about the products you use. That is why it is our duty and responsibility to select quality ingredients only, and which comply with the strictest regulations. We meticulously choose them to offer you products that meet the very highest quality and performance standards.

All the ingredients used in L'Oréal's products are safe and have been subject to a rigorous scientific evaluation of their safety, by our internal experts and also independent experts.

We use an increasing variety of natural ingredients along with synthetic ones as we strive to always be attentive and satisfy all needs and growing expectations in the beauty sector.

Our Commitment to Safety

The safety of our consumers is an absolute priority for us.

Your safety and confidence in using our products is what drives us. Safety assessment is at the heart of what we do. It is a prerequisite for the choice of our ingredients and the conception of any of our products.

We make sure they all comply with the strictest regulations in the world wherever our products are sold.

Ensuring such a guarantee relies on a rigorous system, which encompasses every aspect of our production process worldwide, from the choice of our ingredients and product formulation, to manufacturing and packaging.

Brands

L'Oréal is richly endowed with a portfolio of international brands that is unique in the world and that covers all the lines of cosmetics and responds to the diverse needs of consumers the world over.

1. L'Oréal luxe

L'Oréal Luxe products are available at department stores, cosmetics stores, travel retail, but also own-brand boutiques and dedicated e-commerce websites.

2. Consumer products division

The Consumer Products Division brands are distributed in retail channels.

3. Professional products division

The Professional Products Division distributes its products in salons worldwide.

4. Active cosmetics division

The Active Cosmetics Division brands are sold in healthcare outlets worldwide, including pharmacies, drugstores, and medi-spas.

Ⅳ. Research and Innovation

L'Oréal has always elected to develop strong research because where beauty is concerned, science is essential for innovation. Its Research and Innovation model enables it to respond to the world's vastly diverse beauty expectations.

Research has always been at the heart of L'Oréal's growth, with three major drivers of innovation: active ingredients, formulation, evaluation.

Science, the Driver of Innovation in Cosmetics

For over a century, L'Oréal has built its development on a conviction: only strong research can create cosmetic products that are capable of generating real results. Its Research and Innovation model, unique in the cosmetics industry, is organized around three major entities:

Advanced Research, tasked with continuously enriching scientific knowledge about skin and hair around the world, and discovering new active ingredients;

Applied Research, which develops formulation systems, which are then played out in the different families of products;

Finally, Development, which provides the brands with innovative formulas adapted to their identity and to consumer expectations around the world.

Assets for Innovation

To stay ahead of the game and make major cosmetic innovations available to everyone, the group relies on three major assets. A unique collection of proprietary active ingredients, which advanced research enriches each year with new elements, molecules, or ingredients… Expertise in formulation, this decisive phase that makes it possible to transition from the molecule to the finished product. Every year, thousands of formulas are developed by the teams at L'Oréal. The final asset is its expertise in evaluation, indispensable for bringing new products to the market by demonstrating their safety and effectiveness scientifically and rigorously.

Research That Listens to Consumers

At L'Oréal, innovation has always been nurtured by a constant dialog between science and marketing. It is founded on ever-more precise scientific knowledge of skin and hair around the world, but is also based on attentive listening to consumers on every continent and on the observation of their behavior where beauty is concerned. A true source of inspiration, the great diversity of beauty rituals opens up new fields of exploration.

a Global Vocation

By strengthening its global presence through the six regional poles—Europe, United States, Japan, China, Brazil and India—L'Oréal's Research and Innovation comes closer to its major markets, but also to the wealth of scientific knowledge of each region. In each, the research teams develop strategic partnerships with scientific experts or local start-ups to explore new territories.

Key Words & Phrases 重点词汇

competency[ˈkɒmpɪtənsi] n. 能力；资格
commit[kəˈmɪt] vt. 犯罪；把……交托给；指派……作战；使……承担义务；（公开地）表示意见 vi. 忠于(某个人、机构等)；承诺
infinite [ˈɪnfɪnət] adj. 无限的，无穷的；无数的；极大的
universalise vt. 使一般化，使普遍化，通用化
uniformity[ˌjuːnɪˈfɔːməti] n. 均匀性；一致；同样
reinvent[ˌriːɪnˈvent] vt. 重新使用；彻底改造；重复发明
tailor-made[ˌteɪlər ˈmeɪd] adj. 特制的；裁缝制的
accelerate[əkˈseləreɪt] vt. 使……加快；使……增速 vi. 加速；促进；增加
transformation[ˌtrænsfərˈmeɪʃn] n. [遗]转化；转换；改革；变形
efficacy[ˈefɪkəsi] n. 功效，效力
complementarity[ˌkɒmpləmenˈtærəti] n. 互补性；补充；补足
autonomy[ɔːˈtɒnəmi] n. 自治，自治权

Lesson 2　Brand Introduction of Cosmetics

integrity[ɪnˈtegrəti]　n. 完整；正直；诚实；廉正
synthetic[sɪnˈθetɪk]　adj. 综合的；合成的，人造的　n. 合成物
strive[straɪv]　vi. 努力；奋斗；抗争
attentive[əˈtentɪv]　adj. 注意的；体贴的；留心的
conviction[kənˈvɪkʃn]　n. 定罪；确信；证明有罪；确信，坚定的信仰
demonstrate[ˈdemənstreɪt]　vt. 证明；展示；论证　vi. 示威
territory[ˈterətɔːri]　n. 领土，领域；范围；地域；版图

Practical Activities　实践活动

Part A　Reading comprehension.

1. According to the passage, which of the following is NOT L'Oréal's values and ethical principles? (　　)
A. Passion & Innovation　　　　　　B. Open-mindedness & Responsibility
C. Integrity & Courage　　　　　　　D. Conservative & Casual

2. According to the history of L'Oréal, when was the company founded to become a L'Oréal group? (　　)
A. 1909　　　　B. 1957　　　　C. 1984　　　　D. 2001

3. According to the commitment to safety, L'Oréal relies on a rigorous system, including which of the following? (　　)
A. the choice of ingredients　　　　B. the choice of product formulation
C. manufacturing and packaging　　　D. All of the above

4. How many lines of cosmetics does L'Oréal cover? (　　)
A. Three lines　　B. Four lines　　C. Five lines　　D. Six lines

5. Which of the following brands is NOT L'Oréal Luxe products? (　　)
A. LANCOME　　B. BIOTHERM　　C. Shu uemura　　D. VICHY

6. Where are the Active Cosmetics Division brands sold? (　　)
A. pharmacies　　B. drugstores　　C. medi-spas　　D. all of the above

7. According to Research and Innovation model, what are the three major entities? (　　)
A. Advanced Research　　　　　　　B. Applied Research
C. Development　　　　　　　　　　D. All of the above

Part B　Translation exercise.

1. We have chosen to offer our expertise in the service of women and men worldwide, meeting the infinite diversity of their beauty desires.

2. Always wanting to push back the limits of knowledge means discovering new ways to create products that are truly different and surprising. To always stay one step ahead.

3. As a world leader in beauty, we have, more than others, the duty to preserve the beauty of the planet and to contribute to the well-being of our employees and of the communities in which we are present.

4. With a portfolio of powerful, international brands, L'Oréal enters the 21st century by embracing diversity in its global growth agenda.

5. For over a century, L'Oréal has built its development on a conviction: only strong research can create cosmetic products that are capable of generating real results.

Part C　Brands summary.

How many cosmetic brands are owned by L'oreal Group? Try to summarize them.

全球化妆品行业最有价值品牌50强

知名化妆品品牌中英文对照

Task 2　HERBORIST: A Chinese Premium Brand of Cosmetic

任务二　认识一个中国高档化妆品品牌——佰草集

Herborist or 佰草集（"Bai Cao Ji"）is the only premium brand of Chinese cosmetics which succeeded in its business in Asia and Europe. It's a brand from the Jahwa, a Chinese cosmetics group, created 110 years ago in Shanghai.

The principles of this brand are the Chinese traditional medicine, and the organic cosmetic in Europe.

Ⅰ. Communication

The Chinese brand chose to differentiate its communication per continent. In China, Herborist succeeds in giving an international and contemporary dimension, with a modern logo, a high-end communication, a high-quality packaging, "worthy of the best" (Estee Lauder, Lancôme...).

The logo has a really refined style, the range product is really feminine.

Description of the product:

The Middle Empire reveals its beauty secrets: discover the power of its precious plants. Herborist feeds on ancestral knowledge from the Chinese herbalist to create its exceptional products: a subtle meeting between traditional knowledge and the most innovative formula from contemporary cosmetics. By revealing its massage rituals, Herborist suggests a unique experience of beauty and welfare.

The brand's marketing in China is a success and has been realized by a French agency called Cent Degres.

Ⅱ. Product

The range of products Herborist is not that different from its competitors, presenting face cream and moisturizing face-masks.

III. The Packaging

The real force of this brand is its packaging, really attractive and different.
The "Tai Qi touch" attracts the eye and permits to the product to be different in distribution.
The choice of the color "green" reminds nature but in a modern and innovative way.
Result: the product has a good differentiation from their competitors in sales points.
The package looks like "premium and luxurious", all in surfing on the organic trend.
The national references of each brand, like Shiseido＝Japanese purity, MAC＝American sexy, Lancôme＝French elegance... Herborist will be the natural Chinese medicine. These innovations are in perfect harmony with the market trends for the organic and natural products.
The brand Herborist is cleverly differentiated by integrating the benefits of Chinese traditional herbal medicine, associated with the modern biotechnology.
An image of tradition, of a nature associated with innovation, has permitted to this brand to be well-placed and get a rapid success.

IV. Distribution

Since 2008, the brand is distributed in France in Sephora. Sephora has many shops in the

Magazin Herborist

world and a website of online sales which propose the products of the group Jahwa. In China, there are Herborist's shops and you can find its products in many stores in China, about 800 in the country.

Their products are also presented in the Herborist Spa which was opened in 2012, and propose the products of the brand and unique treatment.

Visit site

V. Price

The price positioning of the Herborist creams and cares for skin is high for its high-end brand image.

VI. The Success of the Brand

The products of the Chinese brand won the consumers' confidence in China and in Occident in 4 years. What a great performance!

Conclusive proves of this success are the increase of the share value of Jahwa. The Chinese cosmetic group is proud of its flagship brand and has launched a perfume, the first Chinese one. The leading cosmetic group in China has other brands less known:

- Chinfie-luxurious cosmetics (online sales)
- GF-cosmetics for men approved by Tony Leung
- Maxam-Skin cares for mature women
- Cocool-Make-up for a younger market

VII. Successful Marketing of a Chinese Brand

In a few years the brand has managed to establish itself in a highly competitive environment with high barriers to entry.

While local competitors have failed to launch a brand Yue Sai as the brand Herborist has to rely on foreign agencies to settle the problem of lack in expertise in branding of luxury in China.

The same problem may result from the case that happens many times that local brands may lack ambition or prefer to use local agencies for their low cost marketing.

Result: they all focusing on the price and eventually find themselves all in dilemma, or further decrease the price, or quit. Good indicators of the success of the brand Herborist are taken as an example in the sphere of marketing in China.

The Herborist is an example of successful premium brand and it is made in China.

Key Words & Phrases 重点词汇

premium[ˈpriːmiəm] n. 额外费用；奖金；保险费；[商]溢价 adj. 高价的；优质的
continent[ˈkɒntɪnənt] n. 大陆，洲，陆地 adj. 自制的，克制的
contemporary[kənˈtempəreri] adj. 发生(属)于同时期的；当代的 n. 同代人，同龄人；同时期的东西
dimension[daɪˈmenʃn, dɪˈmenʃn] n. 方面；[数]维；尺寸；次元；容积
feminine[ˈfemənɪn] adj. 女性的；妇女(似)的；阴性的；娇柔的
precious[ˈpreʃəs] adj. 宝贵的；珍贵的；矫揉造作的
ancestral[ænˈsestrəl] adj. 祖先的；祖传的
ritual[ˈrɪtʃuəl] n. 仪式；惯例；礼制 adj. 仪式的；例行的；礼节性的
differentiate[ˌdɪfəˈrenʃieɪt] vi. 区分，区别
permit[pərˈmɪt] vi. 许可；允许 vt. 许可；允许 n. 许可证，执照
Occident[ˈɒksədənt] n. 西方；欧美国家
flagship[ˈflæɡʃɪp] n. 旗舰；一流；佼佼者
dilemma[dɪˈlemə, daɪˈlemə] n. 困境；进退两难；两刀论法
sphere[sfɪr] n. 范围；球体 vt. 包围；放入球内；使……成球形 adj. 球体的

Practical Activities 实践活动

Group presentation performance task.

Each group is responsible for introducing a famous cosmetics brand.

Make an English PPT and give an oral presentation. Everyone should be involved in your presentation.

知识拓展

The Rebirth of a Chinese Beauty Brand—PECHOIN

Lesson 3 Cosmetic Marketing Strategy
化妆品营销策略

Lead in 课前导入

Thinking and talking:
1. How many types of cosmetic marketing strategies do you know?
2. Which marketing strategy do you think has been mostly used in China? Why?

Related words:

online and offline marketing	线上线下营销	word of mouth marketing	口碑营销
e-commerce platforms	电子商务平台	beauty KOL	美妆达人
internet celebrity economy	网红经济	department stores	百货商店
dedicated counters	专柜	direct selling	直销

Task 1 Cosmetic Online Marketing Strategy: Customization and Personalization

任务一 化妆品网络营销策略——定制化与个性化

The descriptor "Game changer" has been used by many industries. Technology is contributing in making an industry game changer but cosmetics industry has completely altered the traditional form and has evolved as a revolutionary business model innovation.

In the fast moving consumer goods (FMCG) and retail sector, the cosmetics industry is evolving with trends and consumer behavior. Over the last few years, some trends have emerged to redefine the future of beauty.

Customized cosmetics being one of the many trends have been catching the eyes of many customers but the industry is in its nascent stage for entrepreneurs.

The customized cosmetics industry evolved by finding a connection between micro-trends and macro-consumer needs, which are readily shifting. So the real question here: is it justified to invest in something that has become popular due to trend. The industry is creating more micro-trends which are representing serious opportunities.

Here we present, the business model, revenue model, features, and the future of the cosmetics.

Ⅰ. Business Model for Online Customizable Cosmetic Industry

1. Understanding the customers' requirements then creating a product.

The company must preferably work in creating customized cosmetics only.

The customers are given categories from which they choose the product.

After selecting the product, the customer will be asked questions to understand the problem.

Depending on the allergies or reactions, the website selects components to be mixed in the product base.

The brand with the help of logistic partners will create a customized product and deliver it to customers.

2. Providing customers with a base product and making minor changes according to customers.

A company which is already in the cosmetics industry can add a service to increase the

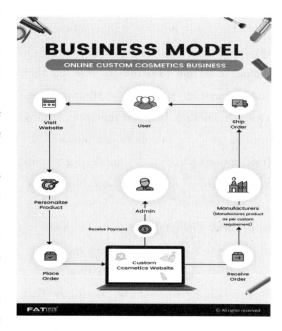

number of customers and engagement.

The customer after choosing the regular product can get the label, fragrance, color, or minor details changed.

The buyer can modify the selections to further customize the base product.

The elements can be predefined by the brand.

With the help of logistic partners, the brand creates and delivers the final product.

II. Feature Analysis of Online Customizable Cosmetic Industry

An algorithm to understand customers' preference Understanding the customer and his/her preference is one of the most important things while customizing a product. The website needs an algorithm to get the specific details of the customers to design the product. While some businesses are using face detection to know the details such as skin tone, others are indulging the customers to choose the best options according to their traits.

Intuitive UI As people have to fill in information about themselves or their preference, they need an intuitive interface which will keep them motivated to take the trouble of filling the information. Business can add gamification to their website to provide the best customer experience.

Specific motive packages There are different types of buyers on a customized cosmetic website. There are a few who need products for specific skin or hair problem. A customer should be allowed to find customized products. For example, hair fall is a common problem, especially for females. The website can provide a package of shampoo, conditioner, mask, serum specifically for hair fall.

Social sharing Social sharing functionality is highly important for customized cosmetics business as customers want to share their created products with their friends or family on social media. This will increase the customer experience and also help promote your business easily.

E-commerce section Businesses can sell their ready-made cosmetics online on their website as not all customers want to customize their cosmetics for their skin.

Review Reviews help build the trust of new customers and improve conversions. Use high-resolution product images to complement the trust-building element of the website.

Gifting options Cosmetics and self-grooming are products which people may want to gift. The website must have an option to order the customized product and also have an option for gift-wrap.

How it works People are not accustomed to the idea of customizing their own cosmetics according to their skin tone, skin type, hair type, etc. A "how it works" section must be there on the website. Make sure that this section is on the homepage or at a prominent page so that people have a fair idea as to how they have to customize the product.

Detailed navigation If you provide customization of many products, make sure you have an easy and well-planned navigation. So that people can easily find the product under the right categories.

Subscription plans　Businesses can tweak their revenue plan and introduce subscription plans. People buy cosmetics on a regular basis, it may be monthly, quarterly or yearly.

FAQ　When it comes to beauty products, customers have a lot of questions. Maintain a web page and publish a list of frequently asked questions (FAQ) to help users find solutions. Here are a few of them:

If I have oily skin should my moisturizer have aloe vera?

What ingredients should be in the shampoo to reduce hair fall?

I am a fair complexion girl. What tone of foundation should I buy?

Ⅲ. Tips for the Success of Customized Cosmetics Business

USP　The business must mention the USP of their customized product on the website. Like some of the important USP, a customized business can showcase is paraben free, chemical free, etc.

Upselling　To increase average order value, custom cosmetics companies have to focus on upselling products. It is easier for them to upsell as when a customer buys a shampoo, he/she also wants to buy a conditioner or serum.

Samples available with a minimum order　To increase sales and revenue of the business, custom cosmetics can provide samples or free products with a minimum order. For example: when a customer orders for $300, they will receive free products with the order. This encourages customers to complete the $300 mark.

Consultation from experts　For most products, customers would know if it is suitable for their skin type or skin tone. The website can have an option where customers can fill in the details about their skin and experts advise them on which product to choose.

Ⅳ. Revenue Model

According to a market analysis, it was analyzed that the global cosmetic products market was valued at USD 532.43 billion in 2017 and is expected to reach a market value of USD 805.61 billion by 2023, registering a CAGR of 7.14% during 2018-2023. The scope of the report is limited to various products, like hair care, skin care, oral care, color cosmetics, fragrances, soaps and shower gels, and sun care products.

Advertisement　The entrepreneur can earn extra revenue by advertising non-competitive businesses on the website. If the website has advertisements for salon services, clothing websites, and more, they can earn some extra revenue.

Subscription　As cosmetics is a consumable product, the customer would want to re-order after a set period of time. To earn continuous revenue, entrepreneurs can introduce a subscription model on their website. Know more on this topic by going through our blog post on subscription box for men & women.

V. What is Online Personalization Cosmetics?

The beauty industry has understood that they have to embrace personalization as no two faces are similar. Famous brands are recognizing this need and implementing personalization. With cosmetics, people are feeling a sense of paralysis and they have a lot of options to choose from. When given a lot of options they are less likely to purchase anything or be happy with the product they have bought. This is the reason beauty brands have started using personalization.

Brands who personalize cosmetics:

CoverGirl　Last year, CoverGirl launched an app, Custom Blend, that analyzes people's skin tone, intensity, and undertone, generating a numeric indicator that matches them with the right products. The whole experience is personalized, right down to the packaging. People can select fonts and label colors, and even include their name on the bottle.

Shiseido　Shiseido has a skincare system that can determine someone's skin texture, pores and moisture content with a photo. It even takes into consideration variables such as temperature and humidity. From there, Optune transmits the data to its IoT-enabled machine, which dispenses the correct serum and moisturizer combination.

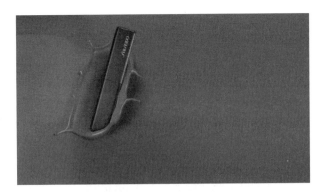

Sephora　Visual Artist is a feature within the Sephora app, a feature that combines facial recognition and augmented reality. It allows consumers to "try on" different products. Skin tones have many subtle variations, so do cosmetics. And those differences don't necessarily translate the way you think they will.

Lesson 3 Cosmetic Marketing Strategy

Ⅵ. What is the Future of the Cosmetics Industry?

The future of the cosmetics industry has taken customization one step forward. First people had to fill in the details and preferences to customize their product. Now there are devices which will create the product for the customers at home. Using water and encapsulated raw materials, ranging from skincare to hygiene and hair products.

In the past few decades, the beauty industry has become bloated and there hasn't been any change in the value chain either. With many newcomers in the online market, people can do much more by deciding what cosmetics they want to buy. Additionally, now they can customize every parameter of the product that suits them personally.

This is indicative of the fact that we have entered an age of e-retail where providing an easy online service isn't enough anymore. It all finally revolves around providing the audience with an individualized experience. Given the fact that there are not many such startups in the cosmetics industry recently, a personalized beauty product service is indeed looking towards a bright future.

Key Words & Phrases 重点词汇

customization[ˈkʌstəmaɪzeɪʃən] n. 定制；用户化；客制化服务
personalization[ˌpəːsənəlaizeiʃn] n. 个性化
alter[ˈɔːltər] vt. 改变，更改 vi. 改变；修改
revolutionary[ˌrevəˈluːʃəneri] adj. 革命的；旋转的；大变革的 n. 革命者
emerge[ɪˈmɜːrdʒ] vi. 浮现；摆脱；暴露
nascent[ˈneɪsnt] adj. 新兴的，初期的，开始存在的，发生中的
evolve[ɪˈvɑːlv] vt. vi. 发展；进化；使逐步形成；推断出
modify[ˈmɒdɪfaɪ] vt. 修改，修饰；更改 vi. 修改
algorithm[ˈælɡərɪðəm] n. 算法，运算法则
intuitive[ɪnˈtuːɪtɪv] adj. 直觉的；凭直觉获知的
UI(user interface) 用户界面
motivate[ˈməʊtɪveɪt] v. 刺激，使有动机，激发……的积极性；成为……的动机；给出理由；申请

gamification[ˌgeɪmɪfɪˈkeɪʃn] n. 游戏化
conversion[kənˈvɜːʃn] n. 转换；变换
prominent[ˈprɒmɪnənt] adj. 突出的，显著的；杰出的；卓越的
navigation[ˌnævɪˈgeɪʃn] n. 航行；航海
subscription[səbˈskrɪpʃn] n. 捐献；订阅；订金；签署
paraben[ˈpærəben] n. 对羟基苯甲酸酯；防腐剂
USP 独特的销售主张
upsell[ˈʌpsel] v. 追加销售
intensity[ɪnˈtensəti] n. 强度；强烈；紧张
visual artist 视觉艺术家
encapsulate[ɪnˈkæpsjuleɪt] vt. 压缩；将……装入胶囊；将……封进内部；概述 vi. 形成胶囊
hygiene[ˈhaɪdʒiːn] n. 卫生；卫生学；保健法
bloat[bləʊt] adj. 肿胀的，鼓起的；饮食过度的，胃胀的 v. 使膨胀，肿胀；腌制；溢出 n. 膨胀；过度，过量；（牛、羊等的）胃气胀

Practical Activities 实践活动

Part A Reading comprehension.

1. According to the passage, which of the following is the business model for Online Customizable Cosmetic Industry? (   )

A. Understanding the customer's requirement then creating a product

B. Providing customers with a base product

C. Making minor changes according to customers

D. All of the above

2. According to the passage, which of following is NOT the meaning of "*An algorithm to understand customers preference*"? (   )

A. There are different types of buyers on a customized cosmetic website

B. The website needs an algorithm to get the specific details of the customers to design the product

C. While some businesses are using face detection to know the details such as skin tone

D. Other businesses are indulging the customers to choose the best options according to their traits

3. According to the passage, which of following can help build the trust of new customers and improve conversions? (   )

A. Social sharing   B. Gifting options   C. Reviews   D. Detailed navigation

4. According to the Advertisement Revenue Model, how can the entrepreneur earn extra revenue? (   )

A. By advertising competitive businesses on the website

B. By advertising non-competitive businesses on the website

C. By introducing a subscription model on their website

D. None of the above
5. Which of the following is NOT the reason why beauty industry has started using personalization? (   )
A. Because no two faces are similar
B. Because people have a lot of options to choose from
C. Because when given a lot of options people are less likely to purchase anything or be happy with the product they have bought
D. Famous brands are recognizing this need and implementing personalization

Part B Translation exercise.

1. In the fast moving consumer goods (FMCG) and retail sector, the cosmetics industry is evolving with trends and consumer behavior. Over the last few years, some trends have emerged to redefine the future of beauty.
2. For example, hair fall is a common problem, especially for females. The website can provide a package of shampoo, conditioner, mask, serum specifically for hair fall.
3. It is easier for them to upsell as when a customer buys a shampoo, he/she also wants to buy conditioner or serum.
4. The website can have an option where customers can fill in the details about their skin and experts advise them on which product to choose.
5. The beauty industry has understood that they have to embrace personalization as no two faces are similar.
6. This is indicative of the fact that we have entered an age of e-retail where providing an easy online service isn't enough anymore. It all finally revolves around providing the audience with an individualized experience.

Task 2 China's Cosmetics Top Brand Marketing Strategies to Unlock the Beauty Market

任务二 解读中国化妆品顶级品牌营销策略

Ⅰ. **The Cosmetics Market in China: a Gold Mine**

Firstly, it is the largest market in the world. Further, it is an extremely dynamic market. Finally, consumers are ready to spend more for higher quality.

Chinese female consumers spend more and more on cosmetics and personal care products and this trend does not seem to be slowing down. China's beauty market is on the way to become the world's largest in terms of sales and market shares by the end of 2020.

How to Explain the Boom of the Cosmetics Market in China?

It is because of the rise of the Chinese middle class and its appetite for all consumer goods, especially beauty brands. Total cosmetics & beauty products revenues retail sales of skin care products in China reached RMB 255 billion in 2018. China is one of the fastest-growing countries among major beauty markets for growth.

II. The Beauty Market Trends in China

1. In Cosmetic: Higher Prices = Better Quality

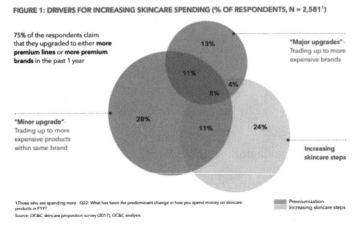

A survey (of OG&C Consultant) explained that 58% of Chinese people claimed that they want a more expensive product, while 36% of Chinese consumers upgraded to more premium brands 39% of Chinese beauty consumers explained that better ingredients and better functionality were the key reasons for their switch to premium brands.

2. Luxury Cosmetics in China on the Rise

For consumers in China, social status is very important. The average Chinese consumer, use luxury brands or expensive products as a way to show off their wealth. Chinese are "new consumers". As a result, they tend to prefer higher-priced items because higher is better Top beauty brands are benefiting from this strong appetite for "cosmetics" and popular products are anti-aging treatments (even to young ladies).

3. International Brand First

For many Chinese girls, foreign brands are "more attractive". There is a kind of "premium selection" that comes with French brands that makes them a favorite among shoppers International branding is the key to catch this segment of the market.

French & European brands are seen as innovative and high end Japanese brands have also a good image in this premium market. Korean brands, other asian brands, and US brands have more mass market positioning.

4. The Problem of Brand-loyalty in China

It is well known that consumers in China tend to have no-loyalty to brands, often trying a different brand However, once they find one they like, they will stick to it.

What it means for you, is that you'll have to spend money to acquire your customers. But once they are converts and are happy about your product, they will come back to it.

At the same time, Chinese female consumers are "very brand-conscious"—buying from "famous brands" as they believe higher prices equate to higher quality. Chinese consumers are now very focused on getting the best product for value and always comparing brands, so consumers' feedback has a huge impact on their decision.

5. Magic Keywords for Cosmetics in China: Natural, Green & Organic

In asia and especially in China, the population is concerned about scandals and pollution Consumers have started to think about the outcome of the pollution on their health and skin This growing awareness has been particularly beneficial to air purifiers companies and skin care brands Meanwhile, the increasing awareness of environmental issues is pushing consumers to turn to more eco-friendly products labeled organic and natural.

6. Cosmeceuticals on the Rise

Cosmeceuticals products are popular because of double functions: beauty and pharmaceutical function such as skin spot treatments, skin brightening, acne treatment, anti-pollution etc.

Chinese consumers are sensitive to products that provide internal benefits. The increasing popularity of multi-functional products like BB Creams and CC Creams is a good example of it.

In 2019, China has voted a law, forbidding to sell cosmetics products as a medical one. Platforms have to enforce this law and brands have to adapt quickly has well Undercover marketing would be the solution here: Kols, Pr, review.

7. Beauty For Men

We can see in China an increase in personal care of young wealthy men, especially in big cities They go to the gym, take care of their appearance and wear luxury clothes.

Brands are creating special line of products targetted at men The men cosmetic segment is estimated to reach 1.9 billion RMB by the end of 2019. Influenced by Korean pop culture, Chinese male consumers are concerned about their their skin. Most male consumers are still quite "conservative" and it is a Niche population that starts to use cosmetics. Here is an example of WeChat Campaign for La Mer China.

8. Beauty Daily Routine Is Changing

Chinese females are using skincare products earlier than the previous generation. Japan and Korea are the kings of it. They are using an average of 21 different products during their skin care routine. Look at some youtube tutorial, that is pretty impressive. The good news for cosmetics brands is that China is taking the same path strongly influenced by its neighbors but also by brands themselves.

Brands have been educating Millennials about the importance of good skin from an early age and an early stage of their cosmetics discovery journey.

Chinese women are also adding more steps to their "daily beauty routine". To the basic function of a facial cleanser and a moisturizing cream, they have added the application of a toner, an eye cream, and a serum.

Around 70%-80% of the beauty consumers also use makeup remover, mask and sun protection cream on a daily basis, which leads to an average of 6-7 steps in their daily skincare routines.

Some parts of the population (beauty "mature" women above 35 years old) were following "up" to 9 steps in their daily skincare routine.

9. The Web: a Huge Influence on Purchase Habits

More than 90% of Chinese girls are influenced by online trends. They are checking beauty news, and actively research new products/reviews.

10. Chinese Girls Are Spending on Skincare Products Earlier

A quarter of Generation Z (born after 1995) are starting to use skincare products as early as 18 years old. 80% of Millennials (born between 1984-1995) started to buy their own skincare products before they were 20 years old.

11. Cosmetics Brands Entering the Chinese Market

You have a cosmetics brand or a beauty brand, and you want to enter the Chinese market. I'm certain you have heard a lot of bad stories. You must already know that the Chinese market is full of opportunities and danger.

To sum up:

China is the most promising market for cosmetics in the world. Above all, it will be the biggest market in the world for almost every beauty product. The demand is increasing. 80% of the market is dominated by international brands. Chinese girls' consuming power increase year after year. Make-up is becoming more and more popular among women.

Ⅲ. The Beauty Market Is Just in Its early stage in China

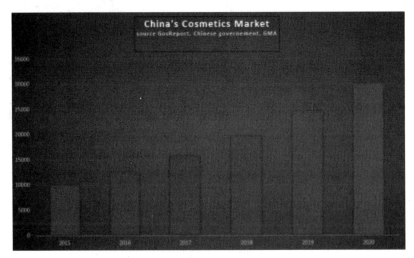

Generally speaking, consumers will spend hours doing researches before purchasing a beauty/skin care product. Chinese consumers spend a lot of time comparing different types of cosmetics and products before taking any decisions.

For instance, more than 90% of girls admitted that they conduct extensive research before they buy skincare products.

There is huge amount of information on skincare brands across multiple channels in China. E-commerce websites such as Tmall and JD are the main channels for browsing across brands and offerings, while social media also plays an essential role in building brand awareness and helping consumers develop product and brand knowledge.

Chinese consumers use a different way to research information than that in the west. Mainly because the tools available in China are different.

Ⅳ. What Are the Top 4 Marketing Keys for Cosmetics in China?

1. Chinese E-commerce Channels

These platforms (like Tmall, Taobao, JD) usually provides a lot of useful information for buyers such as long and detailed product pages.

What consumers really care about is the comments' section. However, most of these comments are not genuine, because of the well-known fact that consumers always check the 3rd party information.

To have a presence on these platforms is an absolute necessity for large brands, while far too expensive for small medium brands. However, Tmall has just released a new service that targets small and medium size brands. It is called TOF or tmall oversea fulfillment.

2. Weibo

Weibo is the number one platform for beauty KOL, cosmetics bloggers. It is the place where all cosmetics superstars start to build their community. Most of them keep publishing on Weibo because it allows them to reach their fans and to get the virality of their posts.

In addition, female consumers usually follow several beauty bloggers because they like tutorials, brand review and they think it is a trustworthy source of information.

Tips for Weibo Marketing

The brand has to have a Weibo:

① Certified Weibo.

② Communication can be focused on the community but should also be commercial (discount, release, coupon etc.).

③ Brands should use sponsored Ads on Weibo. They are efficient and relatively cheap.

④ An effective way to get exposure, and you can target the right users. E.g., female, 25-30, interested in cosmetics.

⑤ Create games. Contest is a good way to engage.

3. Baidu

The Chinese search engine is still very efficient. Why? Because, as we mentioned earlier, consumers are looking for third party information. It is ultra-easy to make research on Baidu. Type your questions and see the results.

Female users will use Baidu to:

① Find a solution to their problems. It is a good way for them to find a product that can solve skin problems, and find the tutorial for makeup. They will actively search for Question & Answer or conversation on forums.

② They will use Baidu to get feedback from users about brands, especially before they decided to buy a product. They will compare different brands and will be influenced by comments or reviews of users for each brand.

They will still consult the brand website. That is a good way to learn more about a brand. But they also know, no brands can be 100% objectives about their products. You would say "My BB cream is great at protecting your skin from UV, however, it feels really sticky". That is exactly the kind of information consumers are looking for, the positive points as well as the negative ones.

Tips for Baidu Marketing

Solely working on your own website and traffic make no sens in China because the user does not trust in corporate communication. They also don't purchase on a brand website but rather on e-commerce platform.

Having a good looking website is still necessary but most of the energy should focus on the Baidu ecosystem.

① Q&A: Baidu Zhidao & Zhihu are good source of information. Get positive feedback from your community is vital for brand reputation. Further, you can help this page rank higher and get maximum exposure.

② Forums: You have a lot of conversations on forums about beauty, and they can rank well if you optimize the search engine visibility.

③ Media: To get your brand's information published on media is a good way to get the trust of potential clients, optimize this publication on Baidu. To let them reach the first page on strategic keywords is a good move to influence Chinese consumers.

Focus your SEO effort on the third-party website rather than on your own website. The goal is to maximize the word of mouth about your brand.

4. Wechat: the Perfect Tool for a Newsletter

Consumers will be influenced by their friends, by the conversation that they can have on WeChat, or discussion that they can follow on groups.

The moment system is the best way for them to get personal feedback about brands, but usually, they do not publish brand reviews.

Groups are a better source of discussion. Peoples discuss brands, solutions to solve their problems or product recommendation.

Some Wechat accounts are also followed by a large number of people.

Wechat Brand Content

Brands should focus on external discussion, brand's mentioned, or the number of shares. Get mentioned by WeChat big accounts is great but, usually, the cost of exposure is extremely high. In other words, we strongly suggest you use this budget in a smarter way.

Community management is the key to Wechat, get in contact with users, engage conversation, create groups per interest. It is time-consuming but also rewarding.

At the moment we speak, advertising on WeChat is expensive. In the future, Wechat will probably develop towards a better Ads system. But keep in mind that the core of Wechat is the users. Users want useful and interesting content.

Key Words & Phrases 重点词汇

mine[maɪn] n. 矿，矿藏；矿山，矿井；地雷，水雷 vt. 开采，采掘；在……布雷 vi. 开矿，采矿；埋设地雷 pron. 我的

social status 社会地位

wealth[welθ] n. 财富；大量；富有

premium['pri:miəm] n. 额外费用；奖金；保险费 adj. 高价的；优质的

loyalty['lɔɪəlti] n. 忠诚；忠心；忠实；忠于……感情

organic[ɔːˈɡænɪk] adj. 有机的；组织的；器官的；根本的

cosmeceutical[kozməˈsjuːtikəl] n. 药妆品

multi-functional adj. 多功能的

tutorial[tjuːˈtɔːriəl] n. (大学导师的) 个别辅导时间，辅导课；教程；辅导材料；使用说明书 adj. 导师的；私人教师的；辅导的

genuine['dʒenjuɪn] adj. 真实的，真正的；诚恳的

Practical Activities 实践活动

Group discussion tasks.

Discuss the following questions with your partners.

1. Which one of the Beauty Market Trends in China is the most outstanding? Why?
2. Which one of the Four Marketing Keys for Cosmetics impress you most? Why?
3. Which is the best distribution way for Cosmetic Companies? Why?

知识拓展

Ten Marketing Tips to Promote Your Cosmetic Brand Effectively

Unit Two
Introduction of Cosmetic Products
化妆品简介

Learning Objectives 学习目标

In this unit you will be able to:
1. Have an overview of cosmetic products' information, such as the label of cosmetic products.
2. Grasp the details about how to introduce and promote the cosmetic products.
3. Practice the activities in relation with introducing and promoting the products.
4. Learn entire cosmetic products by type or usage.

Lesson 4 Skin Care Products
护肤品

 Lead in 课前导入

Thinking and talking:
1. How many kinds of skin care products do you know?
2. How to choose the most suitable skin care products?
3. How to use a variety of skin care products effectively?

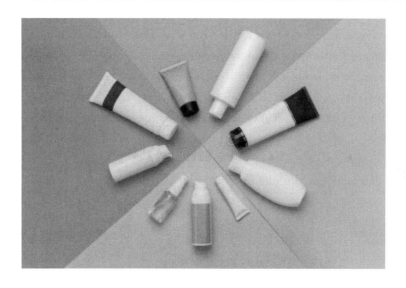

Related words:

skin-care creams	护肤霜	lotion	护肤液；润肤乳
powder	（美容）粉	perfume	香水
lipstick	口红；唇膏	fingernail nail polish	指甲油
permanent waves	卷发；烫发	hair colors	染发剂（彩染）
deodorants	香体剂；除臭剂	bath oils	沐浴油
make-up	化妆品（彩妆）		

Cosmetics are substances used to enhance or protect the appearance or odor of the human body. Cosmetics include skin-care creams, lotions, powders, perfumes, lipsticks, fingernail and toe nail polish, eye and facial makeup, permanent waves related products, colored contact lenses, hair colors related products, hair sprays and gels, deodorants, baby products, bath oils, bubble baths, bath salts, butters and many other types of products. their use is widespread, especially among women in western countries. a subset of cosmetics is called "make up" which refers primarily to colored products intended to alter the user's appearance. many manufacturers distinguish between decorative cosmetics and care cosmetics. there are a few common skin care products below.

Task 1　Cleansers
任务一　认识洗面奶

Cleansing is the most basic and important part of skin care. Only after cleaning the face can you continue the other maintenance steps. Cleansers are products that are intended to clean the skin by removing dirt, oil, makeup, and dead skin cells. They contain special ingredients that help to unclog pores and to prevent skin conditions such as acne. They may also contain moisturizers to prevent the skin from drying out.

Cleansers leave the skin feeling clean and fresh. The safety of cleansers is established by selection of ingredients that are safe and suitable for this purpose. In addition, cleansers are assessed for their potential to cause skin irritation or cause allergic reactions. Product safety is also established though strict adherence to the principles of quality assurance and good manufacturing practices. This includes testing the compatibility of the product with packaging as well as shelf-life stability. Finally, the safety of products is monitored in the marketplace to track any reports of consumer injury. Companies include a phone number on their products where comments or complaints may be reported. A related topic of interest includes Triclocarban.

Key Words & Phrases　重点词汇

allergic reactions　过敏反应
quality assurance　质量保证
good manufacturing practices　良好生产规范
shelf-life　保质期
triclocarban　三氯卡班，$C_{13}H_9Cl_2N_2O$，美国食品药品管理局（FDA）已经禁止在洗手液、香皂等日用品中使用

Practical Activities　实践活动

Part A　Reading comprehension.
1. According to the text, which product has the cleaning effect?（　　）
　A. cleansers　　　B. moisturizers　　　C. creams　　　D. lotions
2. Which one is the first step should we use in skin care?（　　）
　A. cleansers　　　B. moisturizers　　　C. creams　　　D. lotions

Part B True or false questions.

1. Cosmetics include skin-care creams, lotions, make up. (   )
2. Cleansers safety includes testing the compatibility with packaging as well as shelf-life stability. (   )
3. Cleansers can remove dirt, oil, makeup, and dead skin cells from the skin. (   )

Task 2 Moisturizers
任务二 认识保湿霜

Moisturizers are products that are intended to hydrate the skin. Another related topic of interest includes parabens. They also increase the water content of the skin, giving it a smooth appearance. They also provide a barrier against the loss of water from the skin. Moisturizers contain special ingredients that help to replace the oils contained in the skin or to protect against the loss of moisture from the skin.

The safety of moisturizers is established by selection of ingredients that are safe and suitable for this purpose. In addition, moisturizers are assessed for their potential to cause skin irritation or cause allergic reactions. Product safety is also established though strict adherence to the principles of quality assurance and good manufacturing practices. This includes testing the compatibility of the product with packaging as well as shelf-life stability. Finally, the safety of products is monitored in the market-place to track any reports of consumer injury. Companies include a phone number on their products where comments or complaints may be reported.

Key Words & Phrases   重点词汇

moisturizer['mɔɪstʃəraɪzər] *n.* 润肤霜；润肤膏
hydrate['haɪdreɪt] *v.* 使吸入水分；使水合；使成水合物 *n.* 水合物
paraben   *n.* 对羟基苯甲酸酯

Practical Activities   实践活动

Part A Reading comprehension.

1. About moisturizers, which one is not true? (   )
A. Moisturizers can increase the water content of the skin, giving it a smooth appearance.
B. Moisturizers contain special ingredients that help to replace the oils contained in the skin.
C. Moisturizers are assessed for their potential to cause skin irritation or cause allergic reactions.
D. Moisturizers contain special ingredients that help to protect against the loss of oil from the skin.
2. Which skin care products are the most helpful and necessary for people with oily or acne-prone skin? (   )

A. creams B. toners C. moisturizers D. lotions

Part B True or false questions.

Moisturizers safety includes testing the compatibility with packaging as well as shelf-life stability. （   ）

Task 3 Body and Hand Creams and Lotions

任务三 认识霜和乳液

Body and hand creams/lotions are products that are intended to moisturize and soften the body and hands. They are often semi-solid emulsions of oil and water. Body and Hand Creams/Lotions contain special ingredients that help to replace the oils contained in the skin or to protect against the loss of moisture from the skin.

The safety of body and hand creams/lotions is established by selection of ingredients that are safe and suitable for this purpose. In addition, body and hand creams/lotions are assessed for their potential to cause skin irritation or cause allergic reactions. Product safety is also established though strict adherence to the principles of quality assurance and good manufacturing practices.

Key Words & Phrases 重点词汇

semi-solid emulsion 半固体乳液
body and hand creams/lotions 身体和护手霜/乳液

Practical Activities 实践活动

Part A Reading comprehension.

_____ soften the body and hands. （   ）

A. water   B. creams/lotions   C. oil   D. special ingredients

Part B True or false questions.

The safety of body and hand creams/lotions is established by selection of ingredients that are safe and suitable for this purpose. （   ）

Task 4 Toners

任务四 认识爽肤水

Toner looks like water and acts like water. But it's not water. It's packed with so much more than hydrogen and oxygen. Depending on the toner, it also can contain acids, glycerin, antioxidants, and anti-inflammatories.

A toner is a fast-penetrating liquid that delivers skin a quick hit of hydration and helps re-

Lesson 4 Skin Care Products

move some dead cells off the surface of the skin.

As prep for your pores, it brings the skin back to its natural acidic state, sweeping impurities away and helping the skin absorb the skincare products. The skin is like a dried-up sponge. If you put thick cream on a brittle dry sponge, it won't accept it and it isn't prepped for moisture. But if you wet the sponge, the cream will sink in more easily.

Toners are most helpful and necessary for people with oily or acne-prone skin, or for people who want extra cleansing after wearing makeup or other heavy skin products.

That's not all toner can do. There are some additional benefits:

① It shrinks pores. Applying a small amount of toner to a soft cotton ball or pad and gently blotting and wiping your face with it will remove oil and give the appearance of smaller pores.

② It restores your skin's pH balance. Our skin is naturally acidic, typically with a pH balance of between five and six (on a scale from 0 to 14). But that balance can get out of whack after cleansing due to the alkaline nature of soap. When this happens, your skin needs to work overtime to return to its normal levels (and that may result in oil), but using a toner can help restore this balance quickly.

③ It adds a layer of protection. Toners can help close pores and tighten cell gaps after cleansing, reducing the penetration of impurities and environmental contaminants into the skin. It can even protect and remove chlorine and minerals present in tap water.

④ It acts like a moisturizer. Some toners are humectants, which means they help to bind moisture to the skin.

⑤ It refreshes skin. Toner can also be used in lieu of washing your skin when it's oily or dirty. It will leave your skin revitalized even when you're on the go.

⑥ It can prevent ingrown hairs. Toners containing glycolic acid or other alpha hydroxy acids can help to prevent ingrown hairs, so it also aids in grooming.

Key Words & Phrases 重点词汇

hydrogen['haɪdrədʒən] *n.* 氢；氢气
oxygen['ɑːksɪdʒən] *n.* 氧；氧气
acid['æsɪd] *n.* 酸 *adj.* 酸的；酸性的；酸味的
glycerin['ɡlɪsərɪn] *n.* 甘油；丙三醇
antioxidant[ˌæntiˈɑksədənts] *n.* 抗氧化剂
anti-inflammatorie 抗炎
penetrate['penətreɪt] *v.* 穿过；进入；渗透
plump[plʌmp] *adj.* 丰腴的；微胖的
glory['ɡlɔːri] *n.* 荣誉；光荣；灿烂 *v.* 夸耀；得意
sunscreen['sʌnskriːn] *n.* 防晒霜；防晒油
shrink[ʃrɪŋk] *v.* (使)缩水，收缩
whack[wæk] *v.* 猛打；重击；草草放下 *n.* 重击；份儿；一份；量
alkaline['ælkəlaɪn] *adj.* 碱性的；含碱的

chlorine[ˈklɔːriːn]　*n.* 氯；氯气
mineral[ˈmɪnərəl]　*n.* 矿物；矿物质；汽水
humectant[hjuːˈmektənt]　*n.* 保湿剂
lieu[luː]　*n.* 代替；场所，处所
glycolic acid[glaɪˈkɒlək ˈæsɪd]　乙醇酸；羟基乙酸
alpha hydroxy acids　果酸；阿尔法羟基酸
grooming[ˈɡruːmɪŋ]　*n.* 打扮；刷洗；(给动物)梳毛　*v.* (给动物)擦洗，刷洗

Practical Activities　实践活动

Part A　Reading comprehension.

1. When should we use Toners? (　　)
A. Before cleansing　　　　　　　B. Before applying creams
C. After applying creams　　　　　D. After cleansing

2. According to the passage, toner (　　).
A. looks like water　B. is water　　C. acts like oil　　D. isn't a liquid

Part B　True or false questions.

1. Toners can help open pores and tighten cell gaps after cleansing. (　　)
2. A toner helps remove some dead cells off the surface of the skin. (　　)

Task 5　Eye Cream

任务五　认识眼霜

The use of eye cream is a heated debate in the *Allure* office: About half the team believes slathering the undereye area with eye creams can help minimize the appearance of dark circles over time, while the other half seems to think they're moisturizers in tiny tubs and offer nothing more than a placebo effect. Despite all of this, there is one thing we can agree on: No one wants to look like they gotten roughly four hours of sleep. That's why we've combed through dozens of formulas to find the most luxurious and fluffiest formulas—packed with eye-opening ingredients like blood vessel-constricting caffeine and skin-plumping hyaluronic acid—to help fake a full night's rest, no matter how many hours of you have or have not managed to squeeze in.

An eye cream is a specially formulated moisturizer that in most cases has been tested as effective to use near the eyes, and that won't damage the soft tissue around the eyes or cause eye irritation. Many of these creams are made with special ingredients that help either reduce the look of wrinkles around the eyes, provide anti-aging benefits, or help to reduce darker skin tone around the eyes. Some products offer more than one benefit, but all benefit claims have to be taken with a grain of salt, since cosmetic companies are quite well known for making inflated claims about their products.

Some eye cream types contain ingredients that help firm the skin, like caffeine, retinol, or vitamin C. These may temporarily provide a firmer look to the eyes. Those creams that help to lighten skin around the eyes, especially dark circles may contain ingredients such as hydroquinone or Vitamin K. Occasionally you'll find creams that offer anti-aging or firming benefits and reduce dark circles.

Often eye cream formulas have heavier moisturizing ingredients, and some night creams can be used safely around the eyes. You still have to be careful applying any of these creams since even if they don't damage the eyes, they may still hurt if you get some of the cream in your eyes. Eye creams may also be sold as oils or serums instead of creams, which some people find easier to apply.

There is significant variation in price in eye creams, and many come in very small bottles. Trying out a few before buying can help, and sometimes cosmetics companies will offer small samples. Especially when an eye cream is greatly expensive, it's a great idea to see if it works for you, rather than plunk down a lot of money to get a product that doesn't work very well. There are a few eye cream variants that are sold by prescription only. Anything containing hydroquinone may be available in certain countries only by prescription, and some countries ban its use because it may be connected with a higher incidence of skin cancer.

Eye cream with high acid content may make the skin more vulnerable to the sun. If you're using a product with acids, make sure to protect your eyes and skin during the day with a high SPF sunscreen, and with sunglasses. Other eye creams tend to skip compounds that make the skin more vulnerable and lean heavily on moisturizers.

A good way to start choosing an eye cream that will be right for you is to read independent customer reviews of products. You may also get recommendations from friends, or ask your dermatologist what he or she recommends. Aestheticians and make up artists can be another excellent source for discovering which eye creams might truly deliver on their promises.

(Article from https://www.wisegeek.com)

Key Words & Phrases 重点词汇

slather['slæðər] v. 大量地涂抹；厚厚地涂抹
placebo[plə'si:boʊ] n.(给无实际治疗需要者的)安慰剂；(试验药物用的)无效对照剂
fluffy['flʌfi] adj. 绒毛般的；覆有绒毛的；松软的；轻软状
vessel-constricting caffeine 血管收缩咖啡因
skin-plumping 皮肤丰满
hyaluronic acid 透明质酸
anti-aging['ænti 'eɪdʒɪŋ] n. 抗老化；抗衰老
a grain of salt 一粒盐；半信半疑
inflate[ɪn'fleɪt] v. 使充气；膨胀；鼓吹；吹捧
retinol['retɪnɒl] n. 松香油，维生素 A，视黄醇
hydroquinone[ˌhaɪdrəˈkwɪnəʊn] 氢醌；对苯二酚
plunk[plʌŋk] v. 弹拨，刮奏(吉他、琴键等) n.(物体落下)扑嗵声
variant['veəriənt] n. 变种；变体；变形 adj. 变异的；不同的，相异的，不一致的

prescription[prɪˈskrɪpʃn] *n.* 处方；药方；医生开的药
ban[bæn] *v.* 明令禁止；取缔；禁止（某人）做某事（或去某处等） *n.* 禁令
vulnerable[ˈvʌlnərəbl] *adj.*（身体上或感情上）脆弱的，易受……伤害的
dermatologist[ˌdɜːrməˈtɑːlədʒɪst] *n.* 皮肤病医生；皮肤病专家
aesthetician 美学家

Practical Activities 实践活动

Part A Reading comprehension.

1. Eye cream with high_____content may make skin more vulnerable to the sun. （   ）
A. water   B. acid   C. oil   D. salt

2. Eye cream can_____. （   ）
A. help loose the skin
B. decrease darker skin tone around the eyes
C. damage the eyes
D. increase the look of wrinkles around the eyes

Part B True or false questions.

1. Night cream can be used instead of eye cream. （   ）
2. An eye cream is a specially formulated moisturizer. （   ）
3. Eye cream can not help firm the skin. （   ）
4. Eye cream with high acid content will protect the skin from the sun. （   ）
5. A good way to start choosing an eye cream that will be right for you is to read independent customer reviews of products. （   ）

Part C Translate the product labels.

Rare Earth Deep Pore Daily Cleanser	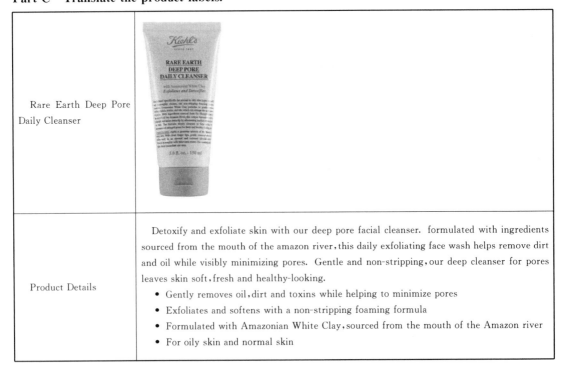
Product Details	Detoxify and exfoliate skin with our deep pore facial cleanser. formulated with ingredients sourced from the mouth of the amazon river, this daily exfoliating face wash helps remove dirt and oil while visibly minimizing pores. Gentle and non-stripping, our deep cleanser for pores leaves skin soft, fresh and healthy-looking. • Gently removes oil, dirt and toxins while helping to minimize pores • Exfoliates and softens with a non-stripping foaming formula • Formulated with Amazonian White Clay, sourced from the mouth of the Amazon river • For oily skin and normal skin

Lesson 4 Skin Care Products

续表

Key Ingredients	Fairly Traded Amazonian White Clay Known as "Magic Earth," Amazonian White Clay has been used for centuries by native Brazilians as a stress relieving, detoxifying therapy. Rich in minerals and extremely porous, Amazonian White Clay is known to help absorb excess oil and detoxify skin. Aloe Barbadensis Aloe Barbadensis has been widely used in traditional and herbal medicine for centuries. Within our formulas, it known to help soothe skin.
Ingredients	Aqua/Water Kaolin Glycerin Alcohol Denat. Sodium Cocoyl Glycinate Coco-Betaine Acrylates Copolymer Sodium Chloride Ci 77891/Titanium Dioxide Solum Diatomeae/Diatomaceous Earth Phenoxyethanol Peg-14M Glycol Distearate Sodium Benzoate Aloe Barbadensis/Aloe Barbadensis Leaf Juice Salicylic Acid Avena Sativa Flour/Oat Kernel Flour Tocopherol Sodium Hydroxide Allantoin
How To Use	Apply a generous amount to damp skin With clean fingertips, gently massage into face and/or neck in upward and outward circular motion Remove thoroughly with lukewarm water Use morning and night Avoid immediate eye area

Rare Earth Deep Pore Cleansing Mask	

Product Details	An efficacious Amazonian White Clay mask that helps minimize pores and detoxify skin. Best for: Oily and normal skin types
Key Ingredients	Amazonian Clay Sourced from the mouth of the Amazon River, Amazonian White Clay is known to help remove excess oil and dead skin that can clog pores. Within our formulas, Amazonian White Clay is known to help detoxify and minimize pores. Aloe Barbadensis Aloe Barbadensis has been widely used in traditional and herbal medicine for centuries. Within our formulas, it known to help hydrate and soothe skin.
How To Apply	After cleansing, apply a thin layer to damp skin, avoiding the immediate eye area Allow to dry for approximately ten minutes When dry, gently remove clay mask using a warm, wet towel and pat dry

知识拓展

Daily Use of Toner

Lesson 5 Color Cosmetics
彩妆产品

Lead in   课前导入

Thinking and talking:
1. Do you wear make up every day?
2. How many kinds of color cosmetic products do you know?
3. Which brand of color cosmetic products is your favorite? Why?

Related words:

facial makeup products	面部彩妆	concealer	遮瑕
lip color	唇妆	foundations	底妆
cheek color	腮红	mascara	睫毛膏
eye makeup remover	眼部卸妆产品	eyeliner	眼线笔

| eye color | 眼妆 | eye shadow | 眼影 |
| eyebrow pencil | 眉笔 | lipstick | 唇膏；口红 |

Task 1　Facial Makeup

任务一　认识脸部化妆品

Ⅰ. Facial Makeup Products

Facial makeup products are products that are used to color and highlight facial features. They can either directly add or alter color or can be applied over a foundation that serves to make the color even and smooth.

Ⅱ. Foundations

Foundations are creams, lotions or powders used as a base for facial makeup. They can be used to color the face and to conceal blemishes to produce an impression of health and youth. Foundations contain ingredients that apply color and/or powder to the skin in a precise and controlled manner. Foundation product safety is established by selection of ingredients that are safe and suitable for this purpose. In addition, foundations are assessed for their potential to cause skin irritation or cause allergic reactions. Product safety is also ensured though strict adherence to the principles of quality assurance and good manufacturing practices. This includes testing the compatibility of the product with packaging as well as shelf-life stability. Finally, the safety of products is monitored in the market-place to track any consumer comments or complaints. Companies include a phone number on their products where comments or complaints may be reported.

III. Concealers

Concealers are products that are intended to conceal or cover up skin imperfections on the face. They typically work by applying color to the skin or through other effects such as the reflection of light or the shininess of the skin. They can be applied using a brush, cloth or spray and are typically used in conjunction with foundations. Concealers contain ingredients that apply color where it is needed in a precise and controlled manner. Concealers safety is established by selection of ingredients that are safe and suitable for this purpose. In addition, concealers are assessed for their potential to cause skin irritation or cause allergic reactions.

IV. Lip Color

Lip colors are products that apply color, texture, and/or shine to the lips using a brush or other applicator. Lip colors contain ingredients that apply color to the lips in a precise and controlled manner. Lip colors can also have multifunctional benefits, such as moisturizing, or may even include sunscreen for SPF protection. Lip color product safety is established by selection of ingredients that are safe and suitable for this intended use and purpose. The colors themselves must be pre-approved by the FDA and are listed in the color additive regulations before they may be used in cosmetics. These color additives are intended for specific use in lip products. Additionally, some colors are also subject to the FDA certification process whereby every manufactured batch must be analyzed and found to meet required purity standards.

V. Face Powders

Face powders are products that are intended to change the appearance of facial skin. They typically work by applying color to the skin or through other effects such as altering the reflection of light or the shininess of the skin. They are typically applied using a brush. Face powders contain ingredients that apply color and/or powder to the skin in a precise and controlled manner.

VI. Cheek Colors

Cheek Colors are products that apply color to accent and highlight the cheekbones or cover up skin imperfections. Cheek color is often applied by women to redden the cheeks so as to provide a more youthful appearance.

Key Words & Phrases   重点词汇

conceal[kən'si:l] v. 隐藏；隐瞒；掩盖
blemish['blemɪʃ] n. 斑点；疤痕；瑕疵 v. 玷污
precise[prɪ'saɪs] adj. 准确的；确切的；精确的；细致的；精细的；认真的；一丝不苟的
irritation[ˌɪrɪ'teɪʃn] n. 生气，气恼；恼人事；烦心事；刺激
shelf-life['ʃelf laɪf] n. 保质期

shininess[ˈʃaɪnɪnɪs]　　*n*. 光泽度；反射；反光度
texture[ˈtekstʃə(r)]　　*n*. 质地；手感
multifunctional[ˌmʌltiˈfʌŋkʃənl]　　*adj*. 多功能的
ingredient[ɪnˈɡriːdiənt]　　*n*. 成分；原料
compatibility[kəmˌpætəˈbɪləti]　　*n*. 和睦相处；兼容性

Practical Activities　实践活动

Part A　Reading comprehension.

1. According to the passage, what's the definition of foundation? (　　)
A. products that apply color, texture, and/or shine to the lips using a brush or other applicator
B. products that are intended to conceal or cover up skin imperfections on the face
C. products include creams, lotions or powders used as a base for facial makeup
D. products that apply color to the eyebrows

2. Regarding the selection of ingredients, what is the most important factor of foundation? (　　)
A. product safety　　B. product package　　C. product price　　D. ingredient types

3. What's the principle of Quality Assurance and Good Manufacturing Practices when it comes to product safety? (　　)
A. the compatibility of the product with packaging as well as shelf-life stability
B. product popularity
C. long shelf-life stability and various packages

4. How to apply concealers? (　　)
A. They can be applied using a cosmetic tissue or pencil
B. They can be applied using a sponge and curler used in conjunction with foundations
C. They can be applied using tweezers especially used in conjunction with foundations
D. They can be applied using a brush, cloth or spray and are typically used in conjunction with foundations

Part B　Translation exercise.

1. Foundations can be used to color the face and to conceal blemishes to produce an impression of health and youth.
2. In addition, foundations are assessed for their potential to cause skin irritation or cause allergic reactions.
3. Concealers typically work by applying color to the skin or through other effects such as the reflection of light or the shininess of the skin.
4. Finally, the safety of products is monitored in the market-place to track any consumer comments or complaints.
5. Lip colors can also have multifunctional benefits, such as moisturizing, or may even include sunscreen for SPF protection.
6. The colors themselves must be pre-approved by FDA and are listed in the color additive

regulations before they may be used in cosmetics.

7. Face powders are products that are intended to change the appearance of facial skin.

Perfect Diary: Strategies for Winning the Chinese Beauty Market

Task 2 Eye Makeup

任务二 认识眼部化妆品

Eye makeup products include products that are used around the eye to enhance the appearance of the eyes and to emphasize the beauty of the eyes. They include such products as eye-shadow, eye liner, eye brow products, and other products that can help to enhance and accent the eyes.

Ⅰ. Eyebrow Pencil

Eyebrow pencils are products that apply color to the eyebrows. They are used to fill in and define the eyebrows. They contain special ingredients that apply color where it is needed in a precise and controlled manner. The products are specially formulated to ensure that potentially harmful microorganisms cannot grow and multiply. The safety of eyebrow pencils is established by selection of ingredients that are safe and suitable for this purpose. In addition, eyebrow pencils are assessed for their potential to cause skin irritation or cause allergic reactions.

Ⅱ. Eye Makeup Remover

Eye Makeup Remover products are intended to help easily remove makeup that has been applied. They help to remove the applied color and to make sure it easily wipes off using a tissue or other cloth.

Ⅲ. Mascara

Mascaras are products intended to enhance the appearance of the eyes by thickening, lengthening, and usually darkening the eyelashes. Mascaras are usually applied with a brush.

Ⅳ. Eyeliner

Eyeliners are products that apply color to the area around the eyes to accent and highlight the appearance of the eyes. Eyeliners are used to emphasize the eyelids and/or to change the perceived shape of the eyes.

Ⅴ. Eye Shadow

Eye shadow is colorful makeup used to accentuate and draw attention to your eyes. It comes in powders, creams, gels and pencil form, and is available in nearly any color you can think of. You can use a single color, or blend two or three shades to give your eye makeup just the look you want.

Key Words & Phrases 重点词汇

enhance[ɪnˈhæns] v. 提高；增强；增进
appearance[əˈpɪərəns] n. 外貌；外观
fill in[fɪl ɪn] 填塞，填平（缝隙或孔洞）
microorganisms[ˌmaɪkrəʊˈɔːɡənɪzmz] n. 微生物；微生物群
mascara[mæˈskɑːrə] n. 睫毛膏 vt. 在……上涂染眉毛油
accent[ˈæksent ˌækˈsent] n. 口音；腔调 v. 着重；强调；突出
perceive[pəˈsiːv] v. 注意到；意识到；认为
accentuate[əkˈsentʃueɪt] v. 着重；强调
blend[blend] v. 使混合 n. 混合品

Practical Activities 实践活动

Part A True or false.

1. The safety of eyebrow pencils is established by selection of pencil shape and size that satisfy different people's habit. (   )
2. Mascara comes in powders, creams, gels and pencil form, and is available in nearly any color you can think of. (   )
3. Eye makeup remover products help to remove the applied color and to make sure it easily wipes off using a tissue or other cloth. (   )
4. Eyebrow pencil can also emphasize the eyelids and/or to change the perceived shape of the eyes. (   )
5. There is no need to concern about that eye makeup products are assessed for their poten-

tial to cause skin irritation or cause allergic reactions. (   )

Part B   Translation exercise.

1. Eye Makeup Products include products that are used around the eye to enhance the appearance of the eyes and to emphasize the beauty of the eyes.

2. The products are specially formulated to ensure that potentially harmful microorganisms cannot grow and multiply.

3. They help to remove the applied color and to make sure it easily wipes off using a tissue or other cloth.

4. Eyeliners are used to emphasize the eyelids and/or to change the perceived shape of the eyes.

5. You can use a single color, or blend two or three shades to give your eye makeup just the look you want.

Part C   Translate the product labels.

Nudes of New York 16 Pan Eyeshadow Palette	
Product Details	This Nudes of New York eyeshadow palette is our first universal palette featuring curated shades that flatter all skin tones as well as every complexion. Creamy formula for texture and color that doesn't look chalky or dull. featuring 16 neutral eyeshadow shades in all of your favorite finishes, from matte eyeshadow to shimmer.
Ingredient	Talc, Caprylic/Capric Triglyceride, Zinc Stearate, Ethylene/Acrylic Acid Copolymer, Boron Nitride, Sodium Dehydroacetate, Phenoxyethanol, Alumina, Polyethylene Terephthalate, Sorbic Acid, Synthetic Fluorphlogopite, Silica, Polyurethane-33, Tin Oxide, [+/-May Contain Mica, CI 77491, CI 77492, CI 77499/Iron Oxides, CI 77891/Titanium Dioxide, CI 77742/Manganese Violet, CI 75470/Carmine, CI 77007/Ultramarines, CI 77400/Bronze Powder, CI 19140/Yellow 5 Lake, CI 77163/Bismuth Oxychloride, CI 19140/Yellow 5, CI 77510/Ferric Ferrocyanide, CI 77000/Aluminum Powder
How to Use/Apply	Step 1. Color the entire eye area. Step 2. Shade the lid. Step 3. Contour the crease. Step 4. Line around eye.

知识拓展

How to Choose Eyeshadow Color?

Task 3　Lip Products

任务三　认识唇部化妆品

Ⅰ. Lipstick Effect Definition

Lipstick effect is a term used to refer to a situation where consumers still manage to buy a luxury product like lipstick, amid an economic crisis. When there is a recession, consumers are expected to focus their purchase on goods that don't impact their already limited income. However, some consumers still have the cash to buy luxury items like premium lipstick.

Ⅱ. A Little More on What is Lipstick Effect

The reason why fast-food restaurants, including movie complexes, perform well during a recession is lipstick effect. Cash-strapped consumers would want to reward themselves with something that helps them to forget their financial problems. People may not afford a holiday in exclusive places, but they can still manage a night out at an affordable place.

Ⅲ. The Psychology Behind Lipstick Effect

The psychology behind the lipstick effect is that amid the economic crisis, most consumers will still afford luxury goods. They are able to fulfill their luxurious needs without straining financially. In the cosmetic market, women would still afford to buy premium lipstick. Other markets like the alcohol market also sell their most expensive drinks during a recession.

The term lipstick effect was coined by Leonard Lauder a former Estee Lauder chairman, following the bursting of the dot-com bubble which sent the U. S. economy into recession in 2000. During the Terrorism attack of 2001 in the United States, Lauder noticed that his company still managed to sell lipstick more than usual. He realized that women resolved to buy products like lipstick in the place of other costlier luxury items. Lauder then concluded that lipstick is a contrary indicator of an economy.

Ⅳ. Pros of the Lipstick Effect

Compared to other economic indicators, the lipstick effect makes sense as it has its basis on economic theory. Lipstick, as well as other small beauty items, is not lesser products. Nonetheless, to consumers, such items are little luxuries they use as a substitute for big luxurious items.

Ⅴ. Cons of Lipstick Effect

One major disadvantage of the lipstick indicator is that lipstick sales' data, including sales for similar products, are not easy for the public to access. For this reason, it is not practically useful to regular investors not unless they are able to track sales from lipstick.

In theory, lipstick sales may be difficult to predict, especially when sales of every item contract simultaneously.

Ⅵ. Lipstick Effect Criticism

A lipstick effect testing was done by economists in 2009 using statistical analysis. A conclusion reached, based on the data collected by Kline & Company, the effect was overestimated.

According to Mintel, a marketing research company, the lip products fell by 3 percent during the great recession. Some economic experts project that the popularity of other products in the cosmetic market is bound to reduce lipstick sales. Amid this shift in the beauty products, the lipstick effect as an economic indicator is slowly losing its relevance.

Key Words & Phrases   重点词汇

recession[rɪˈseʃn]   *n.* 经济衰退；经济萎缩；退后；撤回
premium[ˈpriːmiəm]   *n.* 保险费；额外费用   *adj.* 高昂的；优质的
exclusive[ɪkˈskluːsɪv]   *adj.* 专用的；高档的   *n.* 独家新闻；独家专文；独家报道
strain[ˈstreɪn]   *v.* 损伤；扭伤；尽力
bursting[ˈbɜːstɪŋ]   *adj.* 充满…的；急于（做…）的
costly[ˈkɒstli]   *adj.* 花钱多的；昂贵的
substitute[ˈsʌbstɪtjuːt]   *n.* 代替者   *v.* 取代
simultaneously[ˌsɪməlˈteɪniəsli]   *adv.* 同时；急切地
bound to   一定会；必然；绑定到

Practical Activities   实践活动

Part A   Questions and Answers：
1. What is the definition of Lipstick Effect?

2. Can you explain the reason why people tend to buy lipsticks instead of other products in economic recession?

3. What's the advantage of lipstick effect especially in economic crisis?

4. Why is the performance of lipstick sales difficult to predict?

Part B　Translation Exercise.

1. However, some consumers still have the cash to buy luxury items like premium lipstick.

2. People may not afford a holiday in exclusive places, but they can still manage a night out at an affordable place.

3. The psychology behind the lipstick effect is that amid the economic crisis, most consumers will still afford luxury goods.

4. One major disadvantage of the lipstick indicator is that lipstick sales' data, including sales for similar products, are not easy for the public to access.

5. Nonetheless, to consumers, such items are little luxuries they use as a substitute for big luxurious items.

6. In theory, lipstick sales may be difficult to predict, especially when sales of every item contract simultaneously.

7. A conclusion reached, based on the data collected by Kline & Company, the effect was overestimated.

8. Amid this shift in the beauty products, the lipstick effect as an economic indicator is slowly losing its relevance.

知识拓展

5 Steps to Make Your Lipstick Last Longer

Lesson 6 Hair Products
护发品

Lead in 课前导入

Thinking and talking:
1. How many kinds of hair products do you know?
2. What effect does those hair products?
3. Do you know the purpose of each hair care product?

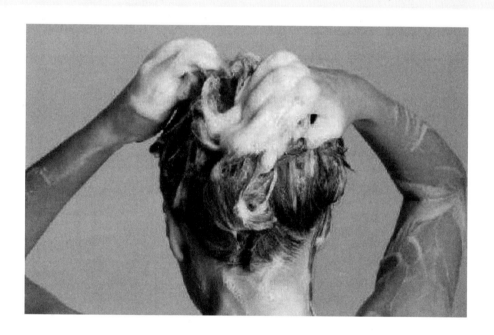

Related words:

hair conditioners	护发素	hair sprays	发胶
hair straighteners and relaxers	直发剂和放松剂	permanent waves	烫发剂
shampoos	洗发水	Rinses	漂洗剂
tonics and dressings	护发液（发油）和敷料	hair bleaches	毛发漂白剂
hair dyes and colors	染发剂和彩染	hair tints	微染剂

Task 1 Hair Care Products

任务一 认识护发产品

Hair care products are those that help to control the properties and behavior of the hair so that it can be maintained in a controlled and desirable manner. This can include hair conditioners, hair sprays, hair straighteners and relaxers, permanent waves, shampoos, rinses, tonics and dressings.

I. Tonics and Dressings

Tonics and dressings are products that are intended to facilitate combing and styling of the hair. They can be applied directly to the hair or by application to the brush or comb. The safety of these products is ensured through the careful selection of ingredients that are safe for these products. Manufacturers conduct extensive safety tests to ensure that these products are safe. Among the tests conducted, are those to ensure that the products are not irritating or that they do not cause allergic reactions.

II. Shampoos and Rinses

Shampoos and rinses are products that are intended to cleanse the hair. Rinses are intended to help condition the hair after shampooing. Some formulations are Shampoos and Rinses in the same product. The safety of these products is ensured through the careful selection of ingredients that are safe for these products. Manufacturers conduct extensive safety tests to ensure that these products are safe. Among the tests conducted are those to ensure that the products are not irritating or that they cause allergic reactions.

III. Permanent Waves

Permanent waves are treatment of the hair intended to produce curls or to alter existing curls. In cosmetology, it is termed a type of curl reformation. Humans have been attempting to add curl to straight hair for thousands of years. The ancient Egyptians did this by ping their hair around wooden sticks; slathering it with mud from hots, letting it bake dry in the sun and then removing the mud. Presumably, the mud had an alkaline chemical makeup that helped to set the curls.

IV. Hair Straighteners and Relaxers

Hair straighteners and relaxers are products that make hair become straight or straighter or to relax tightly curled hair to soften or loosen the curls. The safety of hair straighteners and relaxers is established by selection of ingredients that are safe and suitable for this purpose. In addition, hair straighteners and relaxers are rigorously assessed for their potential to cause scalp and eye irritation. Product safety is also established though strict adherence to the principles of quality assurance and good manufacturing practices.

V. Hair Sprays

Hair spray products are quick-drying liquids sprayed on the hair to keep it in place. They contain ingredients that stick to the hair and hold it in place for a short period of time. The safety of hair sprays is established by selection of ingredients that are safe and suitable for this purpose. In addition, hair sprays are assessed for their potential to cause scalp and eye irritation. Product safety is also established though strict adherence to the principles of quality assurance and good manufacturing practices.

VI. Hair Conditioners

Hair conditioners are used to make hair smooth and silky. They help by replacing materials, such as natural oils, lost during washing. The safety of hair conditioners is established by selection of ingredients that are safe and suitable for this purpose. In addition, hair conditioner products are assessed for their potential to cause scalp and eye irritation. Product safety is also ensured though strict adherence to the principles of quality assurance and good manufacturing practices. This includes testing the compatibility of the product with packaging as well as shelf-life stability.

Key Words & Phrases 重点词汇

rinse[rɪns]   vt.（用清水）冲洗；冲掉…的皂液；漂洗；清洗；（用清水）冲掉   n. 漂洗；冲洗；洗刷；染发剂；漱口液

tonics['tɑːnɪk]   n. 奎宁水，汤力水（一种味微苦、常加于烈性酒中的有气饮料）；补药；滋补品；护发液；护肤液

allergic[əˈlɜːrdʒɪk]   adj.（对…）变态反应的，过敏的；对…十分反感；厌恶

cosmetology[ˌkɑzməˈtɑlədʒi]   n. 美容学，整容术

wrap[ræp]   v. 包，裹（礼物等）；用…缠绕（或围紧）   n.（女用）披肩，围巾；包裹（或包装）材料

slather[ˈslæðər]   v. 大量地涂抹；厚厚地涂抹

alkaline[ˈælkəlaɪn]   adj. 碱性的；含碱的

spray[spreɪ]   n. 浪花；水花；飞沫；喷剂；喷雾的液体；喷雾器   v. 喷；喷洒；向…喷洒

scalp[skælp]   n. 头皮

Practical Activities 实践活动

Part A   Reading Comprehension.

1. According to the text, which product can be used to make hair smooth and silky? (   )
   A. Hair Conditioners   B. Hair dye   C. Hair spray   D. Shampoo

2. What kinds of condition are suited to using tonics and dressing? (   )
   A. that are intended facilitate combing and styling of the hair
   B. that are intended to cleanse the hair
   C. that are intended to produce curls or to alter existing curls
   D. that make hair become straight or straighter or to relax tightly curled hair to soften or loosen the curls

3. What kinds of condition are suited to using Hair Straighteners and Relaxers? (   )
   A. that are intended facilitate combing and styling of the hair
   B. that are intended to cleanse the hair
   C. that are intended to produce curls or to alter existing curls
   D. that make hair become straight or straighter or to relax tightly curled hair to soften or loosen the curls

4. What kinds of condition are suited to using Hair spray? (   )

A. that are intended facilitate combing and styling of the hair

B. that are intended to cleanse the hair

C. that are intended to produce curls or to alter existing curls

D. that to keep the hair in place

Part B　True or False questions

1. They contain ingredients that stick to the hair and hold it in place for a long period of time. (   )

2. Hair Conditioners can help the hair by replacing materials, such as natural oils, lost during washing. (   )

3. Permanent Waves are termed a type of curl reformation. (   )

4. Some formulations are Shampoos and Rinses in the same product. (   )

5. Tonics can not be applied directly to the hair or by application to the brush or comb. (   )

6. Hair Conditioners are used to make hair smooth and silky. (   )

Task 2　Hair Dye and Hair Colour Products

任务二　认识染发产品

Ⅰ. Hair Dye and Hair Coloring Products

Hair dyes and hair colors are products intended to impart color to the hair. Hair dye ingredients have been extensively tested over many years and have been found to be safe. This is established through the testing of individual dying ingredients as well as the other ingredients that they are used to formulate the products. In addition to ensuring the safety of the individual ingredients, hair dyes are assessed to make sure that they are not irritating or cause an unusual incidence of allergic reactions.

Hair color preparations are products that either reduce the color of hair, such as hair bleaches, or products that add color to attain a desirable effect. This can be achieved either with materials that are temporary and only last a few days, to products that last several weeks.

Ⅱ. Hair Bleaches

Hair bleaches are products that lighten the color of the hair by alteration of the coloring components in the hair. They also work through the diffusion of the natural color pigment or artificial color from the hair. The hair bleaching process is central to both permanent hair color and hair lighteners. Hair bleaching ingredients have been extensively tested over many years and have been found to be safe. Their safety is established through the testing of the individual ingredients as well as the other ingredients that are used to formulate the product.

Ⅲ. Hair Tints

Hair tints are products intended to shade the existing color of the hair. Tints may be used also for applying highlighting effects to the hair. Hair tints contain ingredients that have been extensively tested over many years and have been found to be safe. Their safety is established through the testing of individual dying ingredients as well as the other ingredients that are used to formulate the product.

Key Words & Phrases   重点词汇

bleach[ˈbliːtʃɪz]   v. (使)变白，漂白，晒白，褪色

tint[tɪnt]   n. 色调；淡色彩；(一层)淡色，浅色；染发剂；染发   v. 为……轻微染色；给……略微着色；染(发)

Practical Activities   实践活动

Part A   Reading comprehension.

According to the text，which one is not true? (   )

A. Hair bleaches are products that lighten the color of the hair by alteration of the coloring components in the hair.

B. Hair tints are products intended to lighten the existing color of the hair.

C. Hair dyes are products intended to impart color to the hair.

D. Hair colors are products intended to impart color to the hair.

Part B   True or false questions.

1. The Hair bleaching process is central to both permanent hair color and hair lighteners. (   )

2. Hair dye ingredients are safe. (   )

知识拓展

Hair Dye Safety

Lesson 7 Fragrance Product
香水产品

 Lead in 课前导入

Thinking and talking:
1. Do you like to wear perfume?
2. How many perfume brands do you know?
3. What kind of flavor of fragrance product is your favorite?

Related words:

floral	花的，花似的	daisy	雏菊花
fruity	果香味浓的	herbal	药草的；草本的
woody	木质的	delicate	精美的，雅致

blossom	花丛，花簇	aromatic	芳香的，有香味的
top/head notes	前调	middle/heart notes	中调
base notes	尾调		

Task 1 Perfumes

任务一 认识香水

On the face of it, scentless cologne might seem pointless. If it's not going to spritz you with enticing-smelling scent, what's the point?

And yet one perfume company is selling just that-Escentric Molecules fragrances are practically scentless, but apparently mingle with the wearer's natural pheromones to create a unique smell.

After all, no one wants to smell the same as everyone else, do they? It's a scientific new approach to perfumes-there is just one aroma molecule in each fragrance, and the other (odourless) ingredients in the formula are selected to enhance this aroma-molecule and amplify its key qualities.

The concept was created in 2006 by perfumer Geza Schoen, who has been heralded as a rising star in the fragrance industry. The range is now stocked in Selfridges, Liberty London and on Net-a-Porter.

It's a unique departure from traditional perfumes and colognes that are generally made up of top-notes, middle-notes and base-notes, and tend to smell the same on everyone.

Because of something called olfactory adaptation, you never know quite how strongly your fragrance smells—after a few minutes of exposure to a smell, it becomes 80 percent less powerful.

Escentric Molecules are shaking things up though, and celebrities including Rihanna, Beyoncé, Jay-Z and Kate Moss are rumoured to be fans. It likely won't be long before other perfumers are experimenting with the technology too.

The colognes are unisex and smell different on every wearer, but much like the majority of designer perfumes, 100ml of eau de toilette will set you back nearly £70.

New York-based GQ writer Adam Hurly recently tried the fragrances and revealed that he was pleasantly surprised by them, despite being sceptical beforehand: "I liked it. I liked it a lot," he wrote.

Hurly tested two of the Escentric Molecules fragrances: the confusingly-named 'Escentric 01' and 'Molecule 01'. Of the two, he preferred the former.

"Once applied, it reminded me of how my loved ones (and other people I get close enough to smell) have a unique, identifying scent to them—some original musk that permeates their apartment and clothes and car." he explains.

But there's no way of knowing how you'll smell until you spritz yourself with the perfume.

(*Article from* 英语在线翻译网)

Lesson 7 Fragrance Product

Key Words & Phrases 重点词汇

cologne[kəˈləʊn] *n.* 科隆香水，古龙水
spritz[sprɪts] *n.* 细的喷流 *v.* 喷
enticing[ɪnˈtaɪsɪŋ] *adj.* 迷人的；诱人的
pheromones[ˈferəʊməʊnz] *n.* 外激素，信息素
molecule[ˈmɒlɪkjuːl] *n.* 分子
heralded[ˈherəldɪd] *v.* 预示（herald 过去式/过去分词）
mingle[ˈmɪŋɡl] *vt.* 使混合 *vi.* 混合起来
amplify[ˈæmplɪfaɪ] *vt.* 放大，详述
olfactory[ɒlˈfæktəri] adj. 嗅觉的
adaptation[ˌædæpˈteɪʃn] *n.* 改编
moss[mɒs] *n.* 苔
sceptical[ˈskeptɪkl] *adj.* 怀疑的
rumoured[ˈruːməd] *adj.* 谣传的；传说的
musk[mʌsk] *n.* 麝香
permeate[ˈpɜːmieɪt] *vt. & vi.* 弥漫，遍布；渗入

Practical Activities 实践活动

Part A True or false.

1. According to this passage, scentless perfume product doesn't make any sense. (   )
2. It is difficult for Escentric Molecules fragrances to blend the wearer's natural pheromones to create a unique smell. (   )
3. The other (odourless) ingredients in the formula are selected to lessen the aroma-molecule and reduce its key qualities.
4. Because of the olfactory adaptation, people hardly realize how strongly their fragrance smells after a few minutes. (   )
5. New York-based GQ writer Adam Hurly still felt hesitate to try the fragrances since the original musk is too strong. (   )

Part B Translation exercise.

1. On the face of it, scentless cologne might seem pointless. If it's not going to spritz you with enticing-smelling scent, what's the point?
2. Escentric Molecules fragrances are practically scentless, but apparently mingle with the wearer's natural pheromones to create a unique smell.
3. It's a unique departure from traditional perfumes and colognes that are generally made up of top-, middle-and base-notes, and tend to smell the same on everyone.
4. Because of something called olfactory adaptation, you never know quite how strongly your fragrance smells either-after a few minutes of exposure to a smell, it becomes 80 percent less powerful.

5. The colognes are unisex and smell different on every wearer, but much like the majority of designer perfumes, 100ml of eau de toilette will set you back nearly £70.

6. Once applied, it reminded me of how my loved ones (and other people I get close enough to smell) have a unique, identifying scent to them-some original musk that permeates their apartment and clothes and car.

知识拓展

了解更多香水的知识

Task 2　Essential Oils

任务二　认识精油

Lavender essential oil is obtained mostly from the flowers of the lavender plant. The flowers of lavender are fragrant in nature and have been used for making potpourri for centuries. Traditionally, lavender essential oil has also been used in making perfumes. The oil is very useful in aromatherapy and many aromatic preparations and combinations.

(____1____): Lavender essential oil is known as an excellent remedy for various types of pains including those caused by sore and tense muscles, muscular aches, rheumatism, sprains, backache and lumbago. A regular massage with lavender oil can also provide relief from pain in the joints. A study done on postoperative pain relief showed that combining lavender essential oil vapor into the oxygen significantly reduced the amount of pain experienced, versus those patients only revived with oxygen after a major surgery.

(____2____): Lavender essential oil induces sleep which makes it an alternative treatment for insomnia. Frequent studies on elderly patients have shown an increase in their sleep regularity when their normal sleep medication is replaced with some lavender essential oil being placed on their pillows. It has a relaxing impact on people that it can often replace modern medicine for sleep issues.

(____3____): Lavender essential oil has a calming scent which makes it an excellent tonic for the nerves and anxiety issues. Therefore, it can also be helpful in treating migraines, headaches, depression, nervous tension and emotional stress. The refreshing aroma removes nervous exhaustion and restlessness while also increasing mental activity. It has a well-researched impact on the autonomic nervous system, which is why it is frequently used as a treatment for insomnia and also as a way to regulate heart-rate variability. One study showed that people taking tests showed a significant decrease in mental stress and anxiety, as well as increased cognitive function when they inhaled lavender oil and rosemary oil before taking the test.

(___4___): Lavender essential oil is useful for hair care because it has been shown to be very effective on lice, lice eggs, and nits. Furthermore, lavender essential oil has also been shown to be very helpful in the treatment of hair loss, particularly for patients who suffer from alopecia, an autoimmune disease where the body rejects its own hair follicles. A Scottish study reported that more than 40% of alopecia patients in the study reported an increase in hair growth when they regularly rubbed lavender essential oil into their scalp. Therefore, lavender oil is sometimes recommended as a preventative measure for male pattern baldness.

(___5___): There is a significant research on the effects of lavender, in combination with other essential oils, as a way to prevent the occurrence of breast cancer in mice. However, this could be an indication of an increased chance of lavender battling other carcinogenic effects and the presence of cancer.

(*Article from* 组卷网)

Key Words & Phrases   重点词汇

sore[sɔː(r)]   *adj.*（发炎）疼痛的；酸痛的；气恼；愤慨；愤愤不平   *n.* 痛处；伤处
tense[tens]   *adj.* 神经紧张的；担心的；不能松弛的；令人紧张的；绷紧的；不松弛的
postoperative[ˌpəʊst ˈɒpərətɪv]   *adj.* 手术后的
vapor[ˈveɪpə(r)]   *n.* 蒸汽，水气   *v.* 蒸发
induce[ɪnˈdjuːs]   *v.* 劝说；诱使；引起
exhaustion[ɪɡˈzɔːstʃən]   *n.* 筋疲力尽；枯竭
restlessness[ˈrestləsnəs]   *n.* 烦躁不安
alopecian[ˌæləˈpiːʃə]   *n.* 脱发症；秃头症
preventative[prɪˈventətɪv]   *adj.* 预防性的

Practical Activities   实践活动

Part A   Fill in the blanks with the options.

A. Induces sleep   B. Relieves pain   C. Provide hair care
D. Prevents cancer   E. Reduce mental stress and anxiety

Part B   Translation exercise.

1. Lavender essential oil is obtained mostly from the flowers of the lavender plant.
2. A study on postoperative pain relief showed that combining lavender essential oil vapor with the oxygen significantly reduced the amount of pain experienced.
3. Frequent studies on elderly patients have shown an increase in their sleep regularity when their normal sleep medication is replaced with some lavender essential oil being placed on their pillows.
4. Lavender essential oil has a calming scent which makes it an excellent treatment for the nerve and anxiety issues.
5. The refreshing aroma removes nervous exhaustion and restlessness while also increasing mental activity.

6. Therefore, lavender oil is sometimes recommended as a preventative measure for male pattern hairlessness.

7. There is a significant research on the effects of lavender, in combination with other essential oils, as a way to prevent the occurrence of breast cancer in mice.

Dos and Dosn'ts of Essential Oils

精油中英文对照

Lesson 8 Sunscreen
防晒霜

Lead in 课前导入

Thinking and talking:
1. What are the tips of sun protection safety?
2. How to apply and store sunscreen?
3. What is best protection for infants?
4. Do you know the waterproof sunscreen and how to use it?

Related words:

sunscreen	防晒霜	sunburn	晒伤
skin cancer	皮肤癌	skin aging	皮肤老化
broad spectrum sunscreens	广谱防晒霜	SPF values	防晒值
fair-skinned	白皙的，浅色皮肤	dark-skinned	深色皮肤
waterproof	防水的	body washes	沐浴露

Task 1 Sunscreen: How to Help Protect Your Skin from the Sun

任务一 了解防晒产品如何保护皮肤

As an FDA-regulated product, sunscreens must pass certain tests before they are sold. But how you use this product, and what other protective measures you take, make a difference in how well you are able to protect yourself and your family from sunburn, skin cancer, early skin aging and other risks of overexposure to the sun. Some key sun safety tips include:

- Limit time in the sun, especially between the hours of 10 a.m. and 2 p.m., when the sun's rays are most intense.
- Wear clothing to cover skin exposed to the sun, such as long-sleeved shirts, pants, sunglasses, and broad-brimmed hats.
- Use broad spectrum sunscreens with SPF values of 15 or higher regularly and as directed.
- Reapply sunscreen at least every two hours, and more often if you're sweating or jumping in and out of the water.

How to apply and store sunscreen

- Apply 15 minutes before you go outside. This allows the sunscreen (of SPF 15 or higher) to have enough time to provide the maximum benefit.
- Use enough to cover your entire face and body (avoiding the eyes and mouth). An average-sized adult or child needs at least one ounce of sunscreen (about the amount it takes to fill a shot glass) to evenly cover the body from head to toe.

- Frequently forgotten spots: Ears/Nose/Lips/Back of neck/Hands/Tops of feet/Along the hairline/Areas of the head exposed by balding or thinning hair.
- Know your skin. Fair-skinned people are likely to absorb more solar energy than dark-skinned people under the same conditions.
- Reapply at least every two hours, and more often if you're swimming or sweating.

There's no such thing as waterproof sunscreen

People should also be aware that no sunscreens are "waterproof." All sunscreens can eventually be washed off. Sunscreens labeled" "water resistant" are required to be tested according to the required SPF test procedure. The labels are required to state whether the sunscreen remains effective for 40 minutes or 80 minutes when swimming or sweating, and all sunscreens must provide directions on when to reapply.

Storing your sunscreen

To keep your sunscreen in good condition, the FDA recommends that sunscreen containers should not be exposed to direct sun. Protect the sunscreen by wrapping the containers in towels or keeping them in the shade. Sunscreen containers can also be kept in coolers while outside in the heat for long periods of time. This is why all sunscreen labels must say: "Protect the product in this container from excessive heat and direct sun."

Sunscreens for infants and children

Sunscreens are not recommended for infants. The FDA recommends that infants be kept out of the sun during the hours of 10 a.m. and 2 p.m., and to use protective clothing if they have to be in the sun. Infants are at greater risk than adults of sunscreen side effects, such as a rash. The best protection for infants is to keep them out of the sun entirely. Ask a doctor before applying sunscreen to children under six months of age.

For children over the age of six months, the FDA recommends using sunscreen as directed on the Drug Facts label.

Types of sunscreen

Sunscreen comes in many forms, including: lotions, creams, sticks, gels, oils, butters, pastes, sprays.

The directions for using sunscreen products can vary according to their forms. For example, spray sunscreens should never be applied directly to your face. This is just one reason why you should always read the label before using a sunscreen product.

Note: FDA has not authorized the marketing of nonprescription sunscreen products in the form of wipes, powders, body washes, or shampoos.

Key Words & Phrases 重点词汇

ray[reɪ] n. 光线；(热或其他能量的)射线 vt. 放射；照射；(思想；希望等)闪现；发光
long-sleeved shirts 长袖衬衫
sunglasses 太阳镜
broad-brimmed hats 宽边帽
sweat[swet] n. 汗；出汗 v. 出汗；流汗
wrap[ræp] v. 包，裹(礼物等)；用……缠绕(或围紧) n. (女用)披肩，围巾；包裹材料
infants[ˈɪnfənts] n. 婴儿；幼儿
prescription[prɪˈskrɪpʃn] n. 处方；药方

Practical Activities 实践活动

Part A Reading comprehension.

1. Limit time in the sun, especially between the hours of (   ), when the sun's rays are most intense.
   A. 8 a.m. and 10 a.m   B. 10 a.m. and 2 p.m
   C. 12 a.m. and 4 p.m   D. 4 p.m. and 6 p.m

2. How long should you wear sunscreen before going out? (   )
   A. 15 minutes   B. 30minutes   C. 1hour   D. At any time

3. Use enough sunscreen to cover your entire face and body expect (   ).
   A. mouth   B. ears   C. nose   D. hands

4. As of June 2011, SPF values of the sunscreens that pass the broad spectrum test, can indicate (   ).
   A. a sunscreen's UVB protection   B. a sunscreen's UVA protection
   C. a sunscreen's UVA and UVB protection   D. all UV radiation

5. According to the passage, which of the following statement is incorrect. (   )
   A. If your skin is fair, you may want a lower SPF of 15 to 30
   B. To get the most protection out of sunscreen, choose one with an SPF of at least 15
   C. The sun is stronger in the middle of the day compared to early morning and early evening hours
   D. The SPF value indicates the level of sunburn protection provided by the sunscreen product

6. For children over the age of six months, the FDA recommends (   ).
   A. using sunscreen as directed on the Drug Facts label
   B. to ask a doctor before applying sunscreen to children
   C. to use unless absolutely required
   D. to keep them out of the sun entirely

Part B True or false questions.

1. Spray sunscreens can be applied directly to your face. (   )

2. Sunscreen containers should not be exposed to direct sun. (   )
3. Fair-skinned people are likely to absorb more solar energy than dark-skinned people under the same conditions. (   )
4. Sunscreen can be made into lotions, creams, sticks, gels, oils. (   )
5. Sunscreens are made in a broad range of SPF. (   )
6. Sunscreen comes in many forms, including: lotions, creams, sticks, gels, oils, butters, pastes, sprays. (   )

Task 2　Understanding the Sunscreen Label

任务二　认识防晒霜标签

Broad spectrum

Not all sunscreens are broad spectrum, so it is important to look for it on the label. Broad spectrum sunscreen provides protection from the sun's ultraviolet (UV) radiation. There are two types of UV radiation that you need to protect yourself from—UVA and UVB. Broad spectrum provides protection against both by providing a chemical barrier that absorbs or reflects UV radiation before it can damage the skin.

Sunscreens that are not broad spectrum or that lack an SPF of at least 15 must carry the warning:

"Skin Cancer/Skin Aging Alert: Spending time in the sun increases your risk of skin cancer and early skin aging. This product has been shown only to help prevent sunburn, not skin cancer or early skin aging."

Sun protection factor (SPF)

Sunscreens are made in a wide range of SPFs.

The SPF value indicates the level of sunburn protection provided by the sunscreen product. All sunscreens are tested to measure the amount of UV radiation exposure it takes to cause sunburn when using a sunscreen compared to how much UV exposure it takes to cause a sunburn when not using a sunscreen. The product is then labeled with the appropriate SPF value. Higher SPF values (up to 50) provide greater sunburn protection. Because SPF values are determined from a test that measures protection against sunburn caused by UVB radiation, SPF values only indicate a sunscreen's UVB protection.

As of June 2011, sunscreens that pass the broad spectrum test can demonstrate that they also provide UVA protection. Therefore, under the label requirements, for sunscreens labeled "Broad Spectrum SPF [value]", they will indicate protection from both UVA and UVB radiation.

To get the most protection out of sunscreen, choose one with an SPF of at least 15. If your skin is fair, you may want a higher SPF of 30 to 50.

There is a popular misconception that SPF relates to the time of solar exposure. For example, many people believe that, if they normally get sunburned in one hour, then an SPF 15 sunscreen allows them to stay in the sun for 15 hours (e.g., 15 times longer) without getting sunburn. This is not true because SPF is not directly related to the time of solar exposure but to the amount of solar exposure.

The sun is stronger in the middle of the day compared to early morning and early evening hours. That means your risk of sunburn is higher at mid-day. Solar intensity is also related to geographic location, with greater solar intensity occurring at lower latitudes.

Sunscreen ingredients

Every drug has active ingredients and inactive ingredients. In the case of sunscreen, active ingredients are the ones that are protecting your skin from the sun's harmful UV rays. Inactive ingredients are all other ingredients that are not active ingredients, such as water or oil that may be used in formulating sunscreens.

Although the protective action of sunscreen products takes place on the surface of the skin, there is evidence that at least some sunscreen active ingredients may be absorbed through the skin and enter the body. This makes it important to perform studies to determine whether, and to what extent, use of sunscreen products as directed may result in unintended, chronic, systemic exposure to sunscreen active ingredients.

Sunscreen expiration dates

FDA regulations require all sunscreens and other nonprescription drugs to have an expiration date unless stability testing conducted by the manufacturer has shown that the product will remain stable for at least three years. That means, a sunscreen product that doesn't have an expiration date should be considered expired three years after purchase.

To make sure that your sunscreen is providing the sun protection promised in its labeling, the FDA recommends that you do not use sunscreen products that have passed their expiration date (if there is one), or that have no expiration date and were not purchased within the last three years. Expired sunscreens should be discarded because there is no assurance that they remain safe and fully effective.

Lesson 8 Sunscreen

Sunscreens from other countries

In Europe and in some other countries, sunscreens are regulated as cosmetics, not as drugs, and are subject to different marketing requirements. Any sunscreen sold in the United States is regulated as a drug because it makes a drug claim-to help prevent sunburn or to decrease the risks of skin cancer and early skin aging caused by the sun.

If you purchase a sunscreen outside the United States, it is important to read the label to understand the instructions for use and any potential differences between the product and U. S. products.

Key Words & Phrases 重点词汇

ultraviolet[ˌʌltrəˈvaɪələt] *adj.* 紫外线的；利用紫外线的 *n.* 紫外线辐射；紫外光
misconception[ˌmɪskənˈsepʃn] 错误认识；误解
solar[ˈsoʊlər] *adj.* 太阳的；太阳能的
geographic[ˌdʒiəˈɡræfɪk] *adj.* 地理(学上)的；地区(性)的
latitude[ˈlætɪtuːd] *n.* 纬度；纵横；维度
formulate[ˈfɔːrmjuleɪt] *v.* 陈述；制定；制订；形成；阐述
expiration[ˌekspəˈreɪʃn] *n.* 告终；期满；截止

Practical Activities 实践活动

Part A Reading comprehension.

1. Sunscreens that are not broad spectrum or that lack an SPF of at least 15 may been shown only to (   ).
A. help prevent sunburn
B. prevent skin aging
C. prevent skin cancer
D. accelerate to prevent sunburn

2. The SPF value indicates the level of sunburn protection provided by the sunscreen product. SPF is directly to (   ).
A. amount of solar exposure
B. time of solar exposure
C. both UVA and UVB radiation
D. only UVB radiation

3. Why is it important to perform studies to determine whether, and to what extent, use of sunscreen products to sunscreen active ingredients? (   )
A. Some sunscreen active ingredients may be absorbed through the skin and enter the body
B. the protective action of sunscreen products takes place on the surface of the skin
C. water or oil that may be used in formulating sunscreens
D. Every drug has active ingredients and inactive ingredients

Part B True or false questions.

1. Higher SPF values provide greater sunburn protection. (   )
2. Every drug has active ingredients and inactive ingredients. In the case of sunscreen, inactive ingredients are the ones that are protecting your skin from the sun's harmful UV rays. (   )

3. Expired sunscreens should be discarded because there is no assurance that they remain safe and fully effective. (   )

4. Do not use sunscreen products that have passed their expiration date. (   )

5. In Europe and in some other countries, sunscreens are regulated as drugs, not as cosmetics. (   )

6. FDA has authorized the marketing of nonprescription sunscreen products in the form of wipes, towelettes, powders, body washes, or shampoos. (   )

7. The way sunscreen products are used has nothing to do with their form. (   )

8. All sunscreens are broad spectrum. (   )

9. The SPF value indicates the level of sunburn protection provided by the sunscreen product. (   )

10. SPF 15 sunscreen allows people to stay in the sun for 15 hours. (   )

知识拓展

Tips to Stay Safe in the Sun: From Sunscreen to Sunglasses

Unit Three
Cosmetic Ingredients
化妆品原料

Learning Objectives 学习目标

The ingredients in your cosmetic products fall into several different categories, added to provide different characteristics and functions. Each product formula is a careful balance of various ingredients that will work best for what you are trying to use.

In this unit you will be able to:

1. Have an overview of the characteristics and functions of cosmetic ingredients, including cleansing ingredients, skin care ingredients, color additives, auxiliary materials and plant extracts.
2. Learn safety details of the typical cosmetic ingredients.
3. Understand the regulations of color additive.
4. Study more cosmetic ingredients.

Lesson 9 Cleansing Ingredient
清洁类原料

 Lead in 课前导入

Thinking and talking:
1. How do facial cleansers and shampoos achieve cleanness?
2. How many types of cleansing ingredients do you know?
3. How to guarantee the safety of these ingredients?

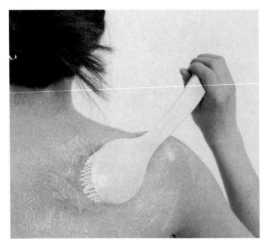
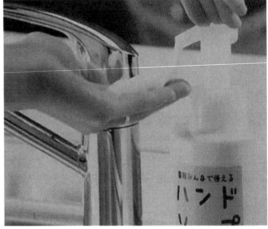

Related words:

facial cleanser	洁面乳	shampoos	洗发水
laundry detergents	洗涤剂	cleaning products	洗涤用品
surfactant	表面活性剂	cosmetic ingredient review (CIR)	化妆品成分评价

Surfactants are the main force in cleaning and therefore make up a big portion of cleaning formulations. Many cleaning products include two or more surfactants in the formula. The choice of surfactants determines where the product will work best.

Task 1　Cocamidopropyl Betaine

任务一　认识椰油酰胺丙基甜菜碱

I. Overview

Cocamidopropyl betaine (CAB) and lauramidopropyl betaine are part of a class of ingredients called amidopropyl betaines. These ingredients consist of various fatty acids bound to amidopropyl betaine.

Cocamidopropyl betaine, lauramidopropyl betaine and related amidopropyl betaines are used mainly as surfactants in cosmetic and personal care products. Surfactants help to clean skin and hair by helping water to mix with oil and dirt so that they can be rinsed away. In this regard, they behave like detergents (soap) and so are found in bath products, skin cleansing products and hair care products, such as shampoos, conditioners and sprays. CAB is also used in household cleaning products, including laundry detergents, hand dishwashing liquids and hard surface cleaners. Other functions reported for these ingredients include: antistatic agent, hair conditioning agent, skin-conditioning agent-miscellaneous, surfactant-cleansing agent, surfactant-foam booster and viscosity increasing agent-aqueous.

II. Safety

The safety of Cocamidopropyl betaine and related amidopropyl betaine ingredients has been assessed on several occasions by the Cosmetic Ingredient Review (CIR) Expert Panel. In 1991, the CIR Expert Panel reviewed the available published scientific literature, and concluded that CAB was safe for use in rinse-off cosmetic products at the levels of use reported in the available literature. Due to the potential for skin irritation at higher use concentrations, however the CIR Expert Panel recommended that for cosmetic products intended to remain on the skin for long periods of time (i. e. , leave-on products), the concentration of CAB should not exceed 3%. Based upon new information showing a substantial increase in the number of uses of CAB, including new uses in aerosol products, and reports of allergic skin reactions in patients who used rinse-off products, the CIR Expert Panel undertook an additional review of CAB and related amidopropyl betaines in 2012. The primary inquiry for the 2012 review related to the presence of the 3,3-dimethylaminopropylamine (DMAPA) and fatty acid amidopropyl dimethylamine (amidoamine) in CAB and other related amidopropyl betaines. DMAPA and amidoamine are present in CAB and other related amidopropyl betaines as manufacturing by-products. DMAPA and amidoamine can sometimes cause allergic skin reactions (i. e. , dermal sensitization). However, the literature reviewed by the CIR Expert Panel has shown that when the levels of DMAPA and amidoamine are reduced, the number of people reacting is also reduced. Based on its review of the published scientific literature, the CIR Expert Panel concluded that cosmetics using CAB and related amidopropyl betaines were safe as long as they were formulated to be non-sensitizing. The CIR Expert Panel also advised industry to continue minimizing the concentration of these manufacturing by-products.

Key Words & Phrases 重点词汇

surfactant[sɜːˈfæktənt] n. 表面活性剂 adj. 表面活性剂的
formulation[ˌfɔːmjuˈleɪʃn] n. 制订，规划；（想法或理论的）系统阐述；表达方式；制剂，配方
betaine[ˈbiːtəiːn] n. 甜菜碱
cocamidopropyl betaine n. 椰油酰胺丙基甜菜碱
lauramidopropyl betaine n. 月桂酰胺丙基甜菜碱
ingredient[ɪnˈɡriːdiənt] n. 原料；要素；组成部分 adj. 构成组成部分的
fatty acid n. 脂肪酸
personal care products n. 个人护理用品
rinse away 冲洗掉
rinse-off products n. 淋洗类产品
detergent[dɪˈtɜːdʒənt] n. 清洁剂；去垢剂
shampoo[ʃæmˈpuː] n. 洗发；洗发精 vt. 洗发
conditioner[kənˈdɪʃənə(r)] n. 护发素
spray[spreɪ] n. 喷雾，喷雾剂；喷雾器；水沫 v. 喷射
household cleaning products n. 家用清洁产品
laundry[ˈlɔːndri] n. 洗衣店，洗衣房；要洗的衣服；洗熨；洗好的衣服
antistatic agent 抗静电剂；静电防止剂
foam booster n. 发泡剂
viscosity increasing agent n. 增稠剂
Cosmetic Ingredient Review(CIR) n. 化妆品成分评估
skin irritation n. 皮肤刺激性
concentration[ˌkɒnsnˈtreɪʃn] n. 浓度
leave-on products n. 驻留类产品
aerosol products n. 气溶胶产品
by-products n. 副产物
allergic[əˈlɜːdʒɪk] adj. 对……过敏的；对……极讨厌的
dermal sensitization n. 皮肤致敏
minimize[ˈmɪnɪmaɪz] vt. 使减到最少；小看，极度轻视 vi. 最小化

Practical Activities 实践活动

Part A Reading comprehension.
1. The safety of cosmetic ingredients should be assessed by the (　　) Expert Panel.
    A. FDA   B. CIR   C. GRAS   D. CBA
2. If a cosmetic products intended to remain on the skin for long periods of time, it is

a (　　).

A. detergent	B. rinse-off product
C. leave-on product	D. foam booster

3. The concentration of CAB in leave-on products should not exceed (　　).

A. 1%　　　B. 2%　　　C. 3%　　　D. 5%

4. Which of the following statement related to CAB is incorrect? (　　)

A. CAB was safe for use in rinse-off cosmetic products at the levels of use reported in the available literature.
B. The concentration of CAB in leave-on products should exceed 3%
C. Cosmetics using CAB were safe as long as they were formulated to be non-sensitizing
D. CAB is a kind of surfactants widely used in cosmetic and personal care products

5. Which of the following statement related to allergic skin reactions is incorrect? (　　)

A. Allergic skin reactions is also called dermal sensitization
B. DMAPA and amidoamine can sometimes cause allergic skin reactions
C. CAB itself can sometimes cause allergic skin reactions
D. By-products in CAB can sometimes cause allergic skin reactions

Part B　Function summary.

What are the functions of cocamidopropyl betaine and related amidopropyl betaines? Try to summarize them.

The Chemistry of Cleaning

Task 2　Other Cleansing Ingredients

任务二　认识其他常见清洁类原料

Ⅰ. Sodium Lauryl Sulfate and Ammonium Lauryl Sulfate

Alkyl sulfate (AS) is an anionic surfactant, usually a sodium salt, derived from fatty alcohol. Alkyl sulfates are high sudsing surfactants. They have been an ingredient in built all purpose granular detergents for many years; today they are more often found in cosmetic products, such as shampoos. Because they are sensitive to water hardness, they perform best

in all purpose detergents that are fully built to inactivate the hardness. Typical alkyl sulfates include sodium lauryl sulfate and ammonium lauryl sulfate.

Sodium lauryl sulfate (SLS) and ammonium lauryl sulfate are widely used surfactant in shampoos, bath products, hair colorings, facial makeup, deodorants, perfumes, and shaving preparations. However, they can also be found in other product formulations.

The safety of sodium lauryl sulfate and ammonium lauryl sulfate has been assessed by the Cosmetic Ingredient Review (CIR) Expert Panel on two separate occasions (1983 and 2002), concluding each time that the data showed these ingredients were safe in formulations designed for brief, discontinuous use, followed by thorough rinsing from the surface of the skin. In products intended for prolonged contact with skin, concentrations should not exceed 1%.

II. Sodium Laureth Sulfate and Ammonium Laureth Sulfate

Ethoxylated alcohol salts are ingredients used primarily in cleansing products, including bubble baths, bath soaps and shampoos. Examples include ammonium myreth sulfate, sodium coceth sulfate, sodium laneth sulfate and sodium myreth sulfate.

In 2010, the Cosmetic Ingredient Review (CIR) Expert Panel evaluated the scientific data for these ingredients and the closely related sodium laureth sulfate (SLES) and ammonium laureth sulfate and concluded that all of these ingredients were safe for use in cosmetics and personal care products in the present practices of use and concentration, when formulated to be non-irritating.

Sodium lauryl sulfate (SLS), also known as sodium dodecyl sulfate and sodium laureth sulfate (SLES) are both surfactants. Both SLS and SLES are very effective ingredients used in cleansing products. In this function, surfactants wet body surfaces, emulsify or solubilize oils, and suspend soil. These ingredients contribute foaming and lathering properties to cleansing products and bubble baths.

There is an e-mail that has been circulating on the internet for several years which falsely states that SLS and SLES, ingredients used primarily in some cosmetic "rinse off" products, can cause cancer. This allegation is unsubstantiated and false, and typical for Internet rumors notorious for publishing inaccurate and untrue information.

Both SLS and SLES are safe for use in cosmetic products. Both ingredients were reviewed in 1983 and re-reviewed in 2002 by the Cosmetic Ingredient Review (CIR) Expert Panel and found to be safe for use in cosmetic and personal care products. SLS and SLES can cause skin irritation in some persons, which is one reason why it is important to follow the label instructions when using a cosmetic product. Complete reports on both ingredients are available from CIR. Substances known to be carcinogenic have been classified and registered by international organizations, such as the World Health Organization or the International Agency for the Research of Cancer as well as the US Environment Protection Agency and the European Union. None of these organizations have classified SLES and SLS as carcinogens. There is no direct or circumstantial evidence that these two ingredients have any carcinogenic potential. The studies that have been conducted on SLS and SLES indicate that both are safe under proper conditions of use.

Ⅲ. Alkylbenzene Sulfonate and Linear Alkylbenzene Sulfonate

Alkylbenzene sulfonate (ABS): A major class of alkyl aryl sulfonate surfactants used in detergents; usually a sodium salt. ABS is anionic and high sudsing. Prior to the mid-1960s, the form of ABS most widely used in detergent formulations had branched hydrocarbon chains, which resisted biodegradation. In 1965, detergent manufacturers voluntarily replaced ABS nationally in household laundry product with a more rapidly biodegradable variety of ABS called linear alkylate sulfonate, or LAS.

Linear alkylbenzene sulfonate (LAS): A readily biodegradable form of alkylbenzene sulfonate surfactant. This is the workhorse of the detergent industry, with sodium dodecyl benzene sulfonate being the most important single type. It is distinguished from an earlier form of alkylbenzene sulfonate, termed ABS, by its linear (straight chain) structure, which provides its good biodegradation properties. All LAS surfactants are anionic and high sudsing, but their sudsing may be controlled by formulation.

Ⅳ. Quaternary Ammonium Compounds

Substances derived from the ammonium cation (NH_4^+) with one or more hydrogen atoms being replaced by organic groups, and for most purposes prepared as a salt (chloride, bromide, sulfate). The nature of the organic groups determines the character and properties of any given quaternary ammonium compound. Some possess disinfecting and deodorizing capabilities and are used in hard surface cleaners. If more than one of the organic groups is fatty in nature, the quaternary ammonium compound is usually water insoluble. Some of these compounds (such as ditallow dimethyl ammonium chloride) are used as the fabric softening and static control agent in rinse-added fabric softeners and detergent-containing fabric softener. If at least one of the organic groups is fatty in nature, the quaternary ammonium compound is a cationic surfactant.

Key Words & Phrases 重点词汇

Ⅰ. Sodium Lauryl Sulfate and ammonium lauryl sulfate

alkyl ['ælˈkɪl] *adj.* 烷基的，烃基的 *n.* 烷基，烃基
alkyl sulfate *n.* 烷基硫酸盐
anionic [ˌænaɪˈɒnɪk] *adj.* 阴离子的
all purpose *n.* 通用的，万能的

granular[ˈɡrænjələ(r)]　adj．颗粒的；粒状的
water hardness　n．水的硬度
sodium lauryl sulfate　n．月桂醇硫酸酯钠
ammonium lauryl sulfate　n．月桂醇聚醚硫酸酯铵
bath product　n．卫浴产品
hair coloring　n．染发
deodorant[diːˈəʊdərənt]　n．除臭剂，体香剂　adj．除臭的，防臭的
shaving[ˈʃeɪvɪŋ]　v．修面，剃（shave 的现在分词）　n．刨花；刮胡子；削

Ⅱ. Sodium Laureth Sulfate and Ammonium Laureth Sulfate

ethoxylated alcohol salts　n．乙氧基醇盐
bubble bath　n．泡沫浴；泡泡浴
sodium laureth sulfate　n．月桂醇聚醚硫酸酯钠
ammonium laureth sulfate　n．月桂醇聚醚硫酸酯铵
irritating[ˈɪrɪteɪtɪŋ]　adj．刺激的；使愤怒的　v．刺激；激怒；使烦恼；使发炎，使不适
non-irritating　n．无刺激性的
emulsify[ɪˈmʌlsɪfaɪ]　vt．使……乳化　vi．乳化
solubilize[ˈsɒljəbəˌlaɪz]　vt．使溶解；使增溶　vi．溶解
suspend[səˈspend]　vt．延缓，推迟；使暂停；使悬浮　vi．悬浮；禁赛
soil[sɔɪl]　n．土地；土壤；泥土　vt．弄脏；污辱　vi．变脏
lather[ˈlɑːðə(r)]　n．肥皂泡；激动　vt．涂以肥皂泡；使紧张；狠狠地打　vi．起泡沫
allegation　n．指控；陈述，主张；宣称；陈词
unsubstantiated[ˌʌnsəbˈstænʃieɪtɪd]　adj．未经证实的，无事实根据的
rumor[ˈruːmə]　n．谣言；传闻　vt．谣传；传说
notorious[nəʊˈtɔːriəs]　adj．声名狼藉的，臭名昭著的
label instruction　n．指示标签，标签说明书
carcinogenic[ˌkɑːsɪnəˈdʒenɪk]　adj．致癌的；致癌物的
registered[ˈredʒɪstəd]　adj．注册的；记名的；登记过的
the World Health Organization　n．世界卫生组织
the International Agency for the Research of Cancer　n．国际癌症研究所
the US Environment Protection Agency　n．美国环境保护署
the European Union　n．欧盟
circumstantial[ˌsɜːkəmˈstænʃl]　adj．依照情况的；详细的，详尽的；偶然的
potential[pəˈtenʃl]　adj．潜在的，可能的　n．潜能，可能性；电势

Ⅲ. Alkylbenzene Sulfonate and Linear Alkylbenzene Sulfonate

alkylbenzene sulfonate（ABS）　烷基苯磺酸盐
aryl[ˈæraɪl; -rɪl]　n．芳香基　adj．芳香基的
sulfonate[ˈsʌlfəneɪt]　n．磺化，磺酸盐　v．使……磺化
sodium salt　钠盐

Lesson 9　Cleansing Ingredient

branched[brɑ:ntʃ]　*adj*. 分枝的；枝状的；有枝的
branched hydrocarbon chains　支链烃
biodegradation[ˌbaɪəʊdegrəˈdeɪʃən]　*n*. [生物]生物降解；生物降解作用
voluntarily[ˈvɒləntrəli]　*adv*. 自动地；以自由意志
linear alkylate sulfonate(LAS)　*n*. 线性烷基苯磺酸盐
workhorse[ˈwɜːkhɔːs]　*n*. 做重活的人；驮马；重负荷机器　*adj*. 工作重的；吃苦耐劳的
linear structure　*n*. 线性结构
straight chain　*n*. [有机化学]直链

Ⅳ. Quaternary Ammonium Compounds

quaternary ammonium compounds　*n*. [有机化学]季铵化合物
derived　[dɪˈraɪvd]　*adj*. 导出的；衍生的　*v*. 衍生出，源于；得到，提取；导出
ammonium cation　*n*. 铵离子
organic groups　*n*. 有机基团
chloride[ˈklɔːraɪd]　*n*. 氯化物
bromide[ˈbrəʊmaɪd]　*n*. [无机化学]溴化物
sulfate[ˈsʌlfeɪt]　*n*. [无机化学]硫酸盐　*vt*. 使成硫酸盐
disinfect[ˌdɪsɪnˈfekt]　*vt*. 将……消毒
deodorize[diˈəʊdəraɪz]　*v*. 除去……的臭气；给……防臭
insoluble[ɪnˈsɒljəbl]　*adj*. 不能解决的；[化学]不能溶解的；难以解释的
fabric softening　*n*. 植物柔软剂
static[ˈstætɪk]　*adj*. 静态的；静电的；静力的　*n*. 静电；静电干扰
cationic[ˌkætaɪˈɒnɪk]　*adj*. 阳离子的　*n*. 阳离子

Practical Activities　实践活动

Part A　Reading comprehension.

1. According to the passage, alkyl sulfates are not （　　）.
   A. high sudsing surfactants
   B. able to inactivate the hardness
   C. anionic surfactant
   D. cationic surfactant

2. According to the passage, sodium lauryl sulfate and ammonium lauryl sulfate can be used in （　　） without limitation.
   A. continuous used product
   B. rinse-off product
   C. leave-on product
   D. products intended for prolonged contact with skin

3. Which of the following statement related to sodium laureth sulfate and ammonium laureth sulfate is incorrect? （　　）
   A. Generally, they are safe for use in cosmetics and personal care product
   B. They can be used primarily in bubble baths, bath soaps and shampoos
   C. They contribute foaming and lathering properties

D. They can't use in rinse-off products

4. According to the passage, which of the following statement related to SLS and SLES is correct? (   )

A. They can cause cancer

B. They can cause skin irritation in some persons

C. They have carcinogenic potential

D. They are unsafe for use in cosmetic and personal care products

5. (Multiple Choice) Surfactants are very effective ingredients used in cleansing products by (   ).

A. wet body surfaces   B. emulsify oils

C. solubilize oils   D. suspend soil

6. According to the passage, alkylbenzene Sulfonate (ABS) is not (   ).

A. anionic   B. high sudsing

C. having branched hydrocarbon chains   D. biodegradation

7. According to the passage, linear Alkylbenzene Sulfonate (LAS) is (   ).

A. cationic   B. no sudsing

C. having branched structure   D. biodegradation

8. By changing the chemical structure, quaternary ammonium compounds don't possess (   ) capabilities.

A. disinfecting   B. deodorizing   C. water insoluble   D. anionic

Part B Matching the terms of the following materials.

```
1. 椰油酰胺丙基甜菜碱      1. ABS
2. 月桂醇硫酸酯钠          2. SLS
3. 月桂醇聚醚硫酸酯钠       3. CAB
4. 线性烷基苯磺酸盐         4. SLES
5. 烷基苯磺酸盐            5. LAS
```

Part C Finding the materials.

Find cocamidopropyl betaine, sodium lauryl sulfate, sodium laureth sulfate, alkylbenzene sulfonate, linear alkylbenzene sulfonate and quaternary ammonium compounds from labels of cosmetic products, such as facial cleansers, shampoos, bath products, deodorants and laundry products.

知识拓展

What Is the History of Soap?
And Where Did Cleaning Come From?

Lesson 10 Skin Care Ingredient
护肤类原料

 Lead in 课前导入

Thinking and talking:
1. How many types of skin care products do you know?
2. How can skin care products take care of your skin?
3. Which skin care ingredient are you most familiar with?

Related words:

skin care products	护肤品	cream	面霜
lotion	露，乳液	moisturizer	保湿剂
emollient	润肤剂		

Body and hand creams/lotions are products that are intended to moisturize and soften the body and hands. They are often semi-solid emulsions of oil and water. Body and hand creams/lotions contain special ingredients that help to replace the oils contained in the skin or to protect against the loss of moisture from the skin. Moisturizers are products that are intended to hydrate the skin. They also increase the water content of the skin, giving it a smooth appearance. They also provide a barrier against the loss of water from the skin. Moisturizers contain special ingredients that help to replace the oils contained in the skin or to protect against the loss of moisture from the skin.

Task 1 Palmitates

任务一 认识棕榈酸酯类

Ⅰ. Overview

In cosmetics and personal care products, the palmitates are used in a wide spectrum of products. The palmitate ingredients act as lubricants on the skin's surface, which gives the skin a soft and smooth appearance. The palmitates are esters of palmitic acid and ethylhexyl, cetyl or isopropyl alcohol. Ethylhexyl palmitate, also called octyl palmitate, is a clear, colorless, practically odorless liquid. Cetyl Palmitate is a white, crystalline, wax-like substance. Isopropyl palmitate (IPP) is a colorless, almost odorless, liquid. In cosmetics and personal care products, they all function as skin conditioning agents—emollient. The palmitates are efficient opacifiers in cream and lotion shampoos. Isopropyl palmitate also functions as a binder which is an ingredient added to compounded dry powder mixtures of solids to provide adhesive qualities during and after compression to make tablets or cakes.

Ⅱ. Safety

The safety of the Palmitates has been assessed by the Cosmetic Ingredient Revies (CIR) Expert Panel. The CIR Expert Panel evaluated the scientific data and concluded that ethylhexyl palmitate, cetyl palmitate and isopropyl palmitate were safe as cosmetic ingredients. In 2001, the CIR Expert Panel considered available new data on ethylhexyl, cetyl and isopropyl palmitate and reaffirmed the above conclusion.

The CIR Expert Panel concluded that the palmitates would be expected to be nontoxic in view of their hydrolysis to palmitic acid and to the corresponding alcohols. Dermal acute and subchronic studies with the palmitates did not show evidence of toxicity. Eye irritation tests on the palmitates produced either very slight ocular irritation or none at all. Human skin tests with the palmitates and with products containing the palmitates were reviewed. One of three products containing 40%-50% ethylhexyl palmitate produced mild irritation. Moisturizers containing 2.5%-2.7% cetyl palmitate were minimally irritating and produced no signs of sensitization, phototoxicity or photo-contact allergenicity. A bath oil product containing 45.6% isopropyl palmitate produced no signs of irritation, sensitization, phototoxicity, or photo contact allergenicity. The CIR Expert Panel noted that ethylhexyl palmitate was not tested at concentrations above 50% for skin irritation and no data on sensitization or phototoxicity were available for this ingredient. Clinical data on cetyl palmitate were limited to concentrations of 2.7%. Based on the available data, the CIR Expert Panel concluded that ethylhexyl palmitate, cetyl palmitate, and isopropyl palmitate were safe as cosmetic ingredients. Ethylhexyl, cetyl, and isopropyl palmitate may be used in cosmetics and personal care products marketed in Europe according to the general provisions of the Cosmetics Directive of the European Union provided the fatty acids and fatty alcohols are not of animal origin. Raw materials of animal origin must comply with European Union animal by-products regulations.

Key Words & Phrases　重点词汇

moisturize[ˈmɔɪstʃəraɪz]　*vi*. 增加水分；变潮湿　*vt*. 使增加水分，使湿润
soften[ˈsɒfn]　*vt*. 使温和；使缓和；使变柔软　*vi*. 减轻；变柔和；变柔软
semi-solid　*n*. 半固体的
emulsion[ɪˈmʌlʃn]　*n*. [药]乳剂；[物理化学]乳状液
moisture[ˈmɔɪstʃə(r)]　*n*. 水分；湿度；潮湿；降雨量
moisturizer[ˈmɔɪstʃəraɪzə(r)]　*n*. 润肤膏
hydrate[haɪˈdreɪt]　*n*. 水合物，水化物　*v*. 补充水分；(使)水合
barrier[ˈbæriə(r)]　*n*. 障碍物，屏障；界线　*vt*. 把……关入栅栏
palmitate[ˈpælmɪteɪt]　*n*. [有机化学]棕榈酸酯；棕榈酸盐
lubricant[ˈluːbrɪkənt]　*n*. 润滑剂；润滑油　*adj*. 润滑的
ester　[ˈestə(r)]　*n*. [有机化学]酯
palmitic acid　*n*. [有机化学]棕榈酸
ethylhexyl　*n*. 乙基己基
cetyl[ˈsitl]　*n*. 十六(烷)基；鲸蜡基
isopropyl alcohol　*n*. [有机化学]异丙醇
ethylhexyl palmitate　*n*. 棕榈酸乙基己酯
octyl palmitate　*n*. 棕榈酸辛酯
odorless[ˈəʊdəlɪs]　*adj*. 没有气味的
cetyl palmitate　*n*. 棕榈酸鲸蜡酯

crystalline[ˈkrɪstəlaɪn]　adj. 透明的；水晶般的；水晶制的
wax[wæks]　n. 蜡；蜡状物　vt. 上蜡　vi. 月亮渐满；增大　adj. 蜡制的
substance[ˈsʌbstəns]　n. 物质；实质；资产；主旨
isopropyl palmitate(IPP)　n. 棕榈酸异丙酯
emollient[ɪˈmɒliənt]　adj. 使平静的；润肤的；镇痛的　n. 润肤霜
opacifier[oˈpæsəfaɪə]　n. 遮光剂；[化工] 不透明剂
binder[ˈbaɪndə(r)]　n. 黏合剂；活页夹；装订工；捆缚者；用以绑缚之物
compounded[kəmˈpaʊndɪd]　adj. 复合的，化合的　v. 混合；组成
adhesive[ədˈhiːsɪv]　n. 黏合剂；胶带　adj. 黏合的；黏性的
compression[kəmˈpreʃn]　n. 压缩，浓缩；压榨，压迫
nontoxic[nɒnˈtɒksɪk]　adj. 无毒的
hydrolysis[haɪˈdrɒlɪsɪs]　n. 水解作用
dermal acute and subchronic studies　n. 皮肤急性和亚慢性研究
toxicity[tɒkˈsɪsəti]　n. 毒性
eye irritation tests　n. 眼刺激性试验
ocular[ˈɒkjələ(r)]　adj. 眼睛的；视觉的；目击的　n. 目镜
ocular irritation　n. 眼部过敏
human skin tests　n. 人体皮肤测试
sensitization[ˌsensətaɪˈzeɪʃn]　n. 敏化作用；促进感受性；感光度之增强
phototoxicity[ˌfəʊtəʊtɒkˈsɪsəti]　n. 光毒性
photo-contact allergenicity　n. 光敏性
clinical[ˈklɪnɪkl]　adj. 临床的；诊所的
provision[prəˈvɪʒn]　n. 规定；条款；准备
animal origin　n. 动物来源
regulation[ˌreɡjuˈleɪʃn]　n. 管理；规则；校准　adj. 规定的；平常的

Practical Activities　实践活动

Part A　Reading comprehension.

1. According to the passage, which of the following is the functions of body and hand creams/lotions?（　　）
A. Moisturize the body and hands
B. Soften the body and hands
C. Protect against the loss of moisture from the skin
D. all of the above

2. According to the passage, which of the following is the functions of moisturizers?（　　）
A. They can hydrate the skin.
B. They can give the skin a smooth appearance.
C. They can provide a barrier against the loss of water from the skin.
D. all of the above

3. According to the passage, which of following gives correct apprence of cetyl palmitate?

(　　)

A. clear, colorless, odorless liquid

B. white, crystalline, wax-like substances

C. white, odorless liquid

D. colorless, wax-like substances

4. According to the passage, isopropyl palmitate (IPP) can act as (　　).

A. emollient　　　B. opacifier　　　C. binder　　　D. all of the above

5. After hydrolysis, palmitates will be changed into (　　) and (　　).

A. palmitic acid　　　　　　　　B. the corresponding alcohols

C. palmitic alcohol　　　　　　　D. the corresponding acids

6. Human safety tests include (　　).

A. dermal acute and subchronic studies

B. eye irritation tests

C. phototoxicity or photo-contact allergenicity

D. all of the above

Part B　Fill in the terms of cosmetic ingredient.

1. _____ are products that are intended to hydrate the skin.

2. _____ is an ingredient added to compounded dry powder mixtures of solids to provide adhesive qualities during and after compression to make tablets or cakes.

3. Body and hand creams/lotions are often semi-solid _____ of oil and water.

4. isopropyl palmitate _____ is a colorless, almost odorless, liquid.

5. The CIR Expert Panel concluded that the palmitates would be expected to be _____ in view of their hydrolysis to palmitic acid and to the corresponding alcohols. Dermal acute and subchronic studies with the palmitates did not show evidence of _____.

Task 2　Dimethicone

任务二　认识聚二甲基硅氧烷

1. Overview

Dimethicone (also known as polydimethylsiloxane)-a silicon-based polymer-is a man-made synthetic molecule comprised of repeating units called monomers. Dimethicone is one of the most widely used ingredients in cosmetics and personal care products and can also be found in many cooking oils, processed foods, and fast food items.

According to 2019 data in US FDA's Voluntary Cosmetic Registration Program (VCRP), dimethicone was reported to be used in 12,934 products. This included products for use near the eye, shampoos and conditioners, hair dyes and colors, bath oils, skin care products, bath soaps and detergents, suntan preparations and baby products. Dimethicone works as an anti-foaming agent, skin protectant, skin conditioning agent, and hair conditioning agent. It prevents water loss by forming a barrier on the skin. Like most silicone materials, dimethicone has a unique fluidity that makes it easily spreadable and, when applied to the skin, gives products a smooth and silky feel. It can also help fill in fine lines/wrinkles on the face, giving it a temporary "plump" look.

2. Safety

Dimethicone has undergone extensive review by scientific experts and authorities around the world. In 2003, the US Cosmetic Ingredient Review (CIR) Expert Panel reviewed the available literature and safety data for dimethicone as well as a group of closely related silicon polymers that function primarily as skin and hair conditioning agents and concluded dimethicone is safe as currently used. The Food and Drug Administration (FDA) reviewed the safety of dimethicone and approved its use as a skin protectant in over-the-counter (OTC) drug products at concentrations from 1%~30%. Dimethicone is also an FDA approved food additive (antifoaming agent). Dimethicone is listed on the EU's Inventory of Cosmetic Ingredients (Cosing) and its use as a cosmetic ingredient is not restricted in any way according to the General Provisions of the Cosmetics Regulation of the European Union. The Joint FAO/WHO Expert Committee on Food Additives (JECFA) has established an acceptable daily oral intake level for dimethylpolysiloxane (dimethicone) of 0-1.5mg/kg body weight. The maximum level (1.5 mg/kg body weight) is 100-fold lower than the level that caused no harmful effects in laboratory studies.

The CIR Expert Panel reviewed a group of silicon polymer derivatives, including dimethicone, that is similar in structure, composition and use. The Expert Panel considered it unlikely that any of the silicone polymers would be significantly absorbed into the skin due to the large molecular weight of these polymers. Human clinical and laboratory absorption studies specific to dimethicone reported it was not absorbed following dermal exposure. Laboratory studies supported the safety of dimethicone following single or repeated oral, skin, or inhalation exposures. Laboratory and human clinical studies showed dimethicone was not irritate to the skin and did not cause allergic skin reactions (i.e., was not a skin sensitizer). It was also reported to be mild to minimally irritate to the eyes. In laboratory reproductive and developmental toxicity studies, no adverse findings were reported in dosed pregnant females or their offspring.

Dimethicone was also shown not to cause genetic mutations in multiple laboratory studies (i.e., not genotoxic). In several historical laboratory studies with mice where dimethicone was administered orally or on the skin for a lifetime, there was no evidence of increased tumor incidence (i.e., dimethicone was not carcinogenic). Evaluating all available scientific data, CIR concluded dimethicone (and the other closely related silicon polymers) are safe as currently used in cosmetics and personal care products.

3. Function

Silicone-based conditioners, such as dimethicone, are agents that aid in smoothing the hair cuticle and increasing hair smoothness and luster. Following water rinsing of a shampoo or conditioner, the silicone is left behind as a thin coating over each individual hair shaft to fill in visible defects in the hair cuticle, improving combing ease. Increased hair friction snags the hair as the comb is drawn for grooming purposes, resulting in hair breakage. This is the most common cause of significant hair loss in normal patients, patients with dandruff, and patients afflicted with female pattern hair loss. The compatibility of the hair can be increased by smoothing the cuticle and coating each individual hair shaft with an agent to decrease friction. Silicones fulfill this need.

Key Words & Phrases　重点词汇

dimethicone, polydimethylsiloxane　*n*. 聚二甲基硅氧烷
silicon['sɪlɪkən]　*n*. 硅；硅元素
polymer['pɑːlɪmər]　*n*. 聚合物
synthetic[sɪn'θetɪk]　*adj*. 综合的；合成的，人造的　*n*. 合成物
molecule['mɑːlɪkjuːl]　*n*. 分子；微小颗粒，微粒
repeating units　*n*. 重复单元，重复单位；重复链段
monomer['mɑnəmɚ]　*n*. 单体；单元结构
FDA's voluntary cosmetic registration program(VCRP)　FDA化妆品自愿注册计划
suntan['sʌntæn]　*n*. 晒黑；土黄色军服；棕色
anti-foaming agent　*n*. 防泡剂
skin protectant　*n*. 护肤剂
fluidity[flu'ɪdəti]　*n*. 流变性；流质；易变性
spreadable['spredəbl]　*adj*. 容易被涂开的
fine line　*n*. 细皱纹
wrinkle['rɪŋkl]　*n*. 皱纹
temporary['tempəreri]　*adj*. 暂时的，临时的　*n*. 临时工，临时雇员
plump[plʌmp]　*adj*. 饱满的；胖乎乎的　*v*. (使)饱满而柔软；变圆，长胖；重重地放下　*adv*. 突然(或重重)坠地；直接地　*n*. 突然前冲；重重坠落
extensive[ɪk'stensɪv]　*adj*. 广泛的；大量的；广阔的
primarily[praɪ'merəli]　*adv*. 首先；主要地，根本上
skin protectant　*n*. 护肤剂
over-the-counter(otc)[ˌoʊvər ðə 'kaʊntər]　*adj*. 非处方的
inventory['ɪnvəntɔːri]　*n*. 存货，存货清单；详细目录；财产清册
EU's inventory of cosmetic ingredients(cosing)　欧盟化妆品成分清单
provision[prə'vɪʒn]　*n*. 规定；条款；准备；供应品　*vt*. 供给……食物及必需品
the Joint FAO/WHO Expert Committee on Food Additives(JECFA)　世卫组织食品添加剂联合专家委员会
daily oral intake level　*n*. 每日口服摄入量
clinical['klɪnɪkl]　*adj*. 临床的；诊所的
human clinical studies　*n*. 人体临床研究
dermal['dɚməl]　*adj*. 真皮的；皮肤的
dermal exposure　*n*. 皮肤接触
exposure[ɪk'spoʊʒər]　*n*. 暴露；曝光；揭露；陈列

inhalation[ˌɪnhəˈleɪʃn]　　n．吸入；吸入药剂
adverse[ədˈvɜːrs]　　adj．不利的；相反的；敌对的
dose[dəʊs]　　n．剂量；一剂，一服　　vi．服药　　vt．给药
pregnant[ˈpregnənt]　　adj．怀孕的；富有意义的
offspring[ˈɔːfsprɪŋ]　　n．后代，子孙；产物
genetic mutation　　n．基因突变
genotoxic[ˌdʒenəʊˈtɒksɪk]　　adj．遗传毒性的
administer[ədˈmɪnɪst]　　v．管理；治理（国家）；给予；执行
tumor[ˈtjʊmɚ]　　n．肿瘤；肿块；赘生物
carcinogenic[ˌkɑːrsɪnəˈdʒenɪk]　　adj．致癌的；致癌物的
cuticle[ˈkjuːtɪkl]　　n．角质层；表皮；护膜
luster[ˈlʌstɚ]　　n．[光学]光泽；光彩　　vi．有光泽；发亮　　vt．使有光泽
shaft[ʃæft]　　n．竖井；通风井；（电梯的）升降机井；杆，柄　　vt．欺骗；苛待
friction[ˈfrɪkʃn]　　n．摩擦，摩擦力
snag[snæg]　　n．障碍；意外障碍；突出物　　vt．抓住机会；造成阻碍；清除障碍物　　vi．被绊住；形成障碍
grooming[ˈgruːmɪŋ]　　n．（动物）刷洗，梳毛；梳妆；培养　　v．（动物）刷洗，梳毛；梳妆；培养
breakage[ˈbreɪkɪdʒ]　　n．破坏；破损；裂口；破损量
dandruff[ˈdændrəf]　　n．头皮屑

Practical Activities　实践活动

Part A　Reading comprehension.

1. According to the passage, which of the following is the functions of dimethicone? (　　)
A．anti-foaming agent　　　　B．skin protectant
C．skin and hair conditioning agent　　D．all of the above

2. According to the passage, which of the following is not the feature of dimethicone? (　　)
A．Prevent water loss by forming a barrier on the skin
B．Give products a smooth and silky feel
C．Hard to spreadable
D．Give skin a temporary "plump" look

3. According to JECFA, the level that caused no harmful effects in laboratory studies is (　　).
A．0mg/kg　　　B．1.5mg/kg　　　C．15mg/kg　　　D．150mg/kg

4. It unlikely that any of the silicone polymers would be significantly absorbed into the skin due to (　　).
A．They are easy to spreadable
B．They have large molecular weight
C．They can fill in fine lines/wrinkles

D. They can form a barrier on the skin

5. Allergic skin reaction has the same meaning of (   ).

A. skin sensibility   B. genotoxic

C. carcinogenic   D. dermal exposure

Part B Fill in the terms of safety.

1. Dimethicone was not absorbed following dermal exposure, supported the _____ of dimethicone.

2. Something can cause allergic skin reactions is a _____ .

3. Something can cause genetic mutations is _____ .

4. Something can increased tumor incidence is _____ .

Part C Summary.

What scientific data were considered by CIR to make the conclusion that dimethicone (and the other closely related silicon polymers) are safe as currently used in cosmetics and personal care products? Try to summarize them.

Polymers

Task 3   Glycerin

任务三 认识甘油

1. Overview

Glycerin (sometimes called glycerol) is a naturally occurring alcohol compound found in all animal, plant, and human tissues, including the skin and blood. Glycerin used in cosmetics and personal care products can be obtained from natural sources (e.g., soybeans, cane, or corn syrup sugar) or manufactured synthetically. This synthetic form is chemically identical to naturally-occurring glycerin and the body handles both the same way.

Glycerin is used safely in numerous cosmetics and personal care products such as soaps, toothpaste, shaving cream, and skin/hair care products to provide smoothness and lubrication. It is also a well-known humectant that prevents the loss of moisture from products so they don't dry out as quickly. Other reported functions for glycerin include use as a fragrance ingredient, denaturant, hair conditioning agent, oral care agent, skin conditioning agent—humectant, skin protectant, oral health care drug, and viscosity decreasing agent.

According to 2019 data in US FDA's Voluntary Cosmetic Registration Program (VCRP), glycerin is the third most frequently used ingredient in cosmetics (after water and fragrance). It was reported to be used in 23,366 products. This includes products for use near the eye, lipsticks, hair dyes and colors, bath soaps and detergents, skin care products, suntan preparations, and baby products.

According to a survey conducted by the Personal Care Products Council (PCPC) in 2014, glycerin was used at concentrations up to 99.4% in some skin cleaning products.

2. Safety

Scientific data supporting the safety of glycerin as used in cosmetics and personal care products were thoroughly reviewed in 2014 by the Cosmetic Ingredient Review (CIR) Expert Panel. Based on the available literature and data, the Expert Panel concluded glycerin is safe in the present practices of use and concentration (i.e., up to 79% in leave-on products, 99% in rinse-off products).

The US Food and Drug Administration (FDA) recognizes glycerin as generally recognized as safe (GRAS) for its use in food packaging and it is a multiple-purpose GRAS food substance when used in accordance with good manufacturing practices. Glycerin is on FDA's list of approved direct and indirect food additives.

Glycerin is listed on the EU's Inventory of Cosmetic Ingredients (CosIng) and is not restricted in any way, according to the general provisions of the Cosmetics Regulation of the European Union. Glycerin derived from raw materials of animal origin must comply with European Union animal by-products regulations.

Key Words & Phrases 重点词汇

glycerin[ˈɡlɪsərɪn]　　n. 甘油
glycerol[ˈɡlɪsərɪn]　　n. 甘油
tissue[ˈtɪʃuː]　　n. 纸巾；薄纱；一套　　vt. 饰以薄纱；用化妆纸揩去
soybean[ˈsɔɪˌbin]　　n. 大豆；黄豆
toothpaste[ˈtuːθpeɪst]　　n. 牙膏
shaving cream　　n. 刮胡膏，剃须膏
humectant[hjuːˈmektənt]　　adj. 湿润的；湿润剂的　　n. [助剂]湿润剂
denaturant[dɪˈnetʃərənt]　　n. 变性物质；酒类变性剂
oral care agent　　n. 口腔护理剂
viscosity decreasing agent　　n. 降黏剂
survey[ˈsɜːrveɪ]　　n. 调查；测量；审视　　vt. 调查；勘测；俯瞰　　vi. 测量土地
the personal care products council(PCPC)　　n. 个人护理用品委员会
generally recognized as safe(GRAS)　　n. 通常被认为是安全的物质

Practical Activities 实践活动

Part A　Reading comprehension.

1. According to the passage, which of the following statement is correct? (　　)

A. Glycerin obtained from plants is manufactured glycerin.

B. Natural sourced glycerin is much safer than manufactured glycerin.

C. Human body can't handle glycerin that manufactured synthetically.

Lesson 10 Skin Care Ingredient

D. Both natural sourced glycerin and manufactured glycerin can be used safely in cosmetics product.
2. According to the passage, glycerin can provide (   ) in cosmetic products.
A. smoothness   B. lubrication   C. humectant   D. all of the above
3. According to the passage, the top threey used ingredients in cosmetics are (   ).
A. glycerin   B. water   C. fragrance   D. all of the above
4. Glycerin is safe up to (   ) in leave-on products.
A. 99.4%   B. 79%   C. 99%   D. 56%
5. Glycerin derived from raw materials of animal origin must comply with (   ) if used in cosmetics and personal care products marketed in Europe.
A. US Food and Drug Administration regulations
B. generally recognized as safe lists
C. EU's Inventory of Cosmetic Ingredients
D. European Union animal by-products regulations

Part B Translation Exercise.
1. Glycerin used in cosmetics and personal care products can be obtained from natural sources or manufactured synthetically.
2. It is a well-known humectant that prevents the loss of moisture from products so they don't dry out as quickly.
3. Glycerin is the third most frequently used ingredient in cosmetics (after water and fragrance).
4. Glycerin was used at concentrations up to 99.4% in some skin cleaning products.
5. Glycerin is safe in the present practices of use and concentration (i.e., up to 79% in leave-on products, 99% in rinse-off products).

知识拓展

Hyaluronic Acid

Lesson 11 Color Additives
化妆品颜色添加剂

Lead in 课前导入

Thinking and talking:
1. How many types of makeup products do you know?
2. How many colors can you name?
3. Are you familiar with the regulations of color additives?

Related words:

makeup products	化妆品	color additive	着色剂
safety	安全	regulation	管理
dye	染料	pigment	颜料
lake	色淀	colour index	颜色索引

Task 1 Color Additives Used in Cosmetics

任务一 化妆品中使用的颜色添加剂

Ⅰ. Overview

Facial makeup products are products that are used to color and highlight facial features. They can either directly add or alter the color or can be applied over a foundation that serves to make the color even and smooth.

Color additives provide consumers with everything from the red tint in their blush to the green hue of their mint-flavored toothpaste. They are dyes, pigments, or other substances that can impart color when added or applied to a food, drug or cosmetic. They can be found in a wide range of consumer products—from cough syrup and breakfast cereal to contact lenses and eyeliner. Modern day color additives have been safely used for more than 150 years to make cosmetic and personal care products decorative, attractive and appealing. Color additives may be used to create a product image or recognition, the "mood" of the product, or other visual impressions. Prior to the development of the wide color palette currently available, products tended to appear drab and colors were very unstable and faded quickly. Mixing colors to achieve the exact desired effect requires great skill and knowledge of the properties of the ingredients and products—it is truly an art form.

The visual perception of color occurs primarily by the absorption and/or reflection of visible light by the product and corresponds to humans seeing red, yellow, blue, green, black, etc. Such color derives from the different wavelengths of light interacting with the light receptor cells in the eye and sending a message to the brain. Color additives have long been a part of human culture. Archaeologists date cosmetic colors as far back as 5000 B. C., and ancient Egyptian writings tell of drug colorants. Many of us are familiar with and recognize drawings of Cleopatra with her wonderful eye makeup. Historians say food colors likely emerged around 1500 B. C.

Before the development of modern technology, colors primarily came from substances found in nature, such as indigo, turmeric, paprika and saffron. But as the 19th century approached, new kinds of colors appeared that offered marketers wider coloring possibilities. These colors, made in the laboratory, were found to be much more stable with greater coloring intensity, meaning that less colorant could be used in the product to accomplish the same effect. They also could be produced without using plants harvested in the wild.

II. Safety

Color additives are subject to a strict system of approval under the Federal Food, Drug, and Cosmetic Act (FD & C Act). Except in the case of coal-tar hair dyes, all color additives used in cosmetics must be approved by FDA. Color additive violations are a common reason for detaining imported cosmetic products offered for entry into the United States.

According to FDA regulations, if your product (except coal-tar hair dyes) contains a color additive, you must adhere to requirements for:

- Approval. All color additives used in cosmetics (or any other FDA-regulated product) must be approved by FDA. There must be a regulation specifically addressing a substance's use as a color additive, its specifications, and restrictions.
- Certification. In addition to approval, for a number of color additives, every batch made must be certified by FDA if they are to be used in cosmetics (or any other FDA-regulated product) marketed in US.
- Identity and specifications. All color additives must meet the requirements for identity and specifications stated in the Code of Federal Regulations (CFR).
- Use and restrictions. Color additives may be used only for the intended uses stated in the regulations that pertain to them. The regulations also specify other restrictions for certain colors, such as the maximum permissible concentration in the finished product.

According to Linda Katz, M. D., M. P. H., the director of the Office of Cosmetics and

Colors in FDA's Center for Food Safety and Applied Nutrition, "Color additives are very safe when used properly." The FDA conducts detailed safety reviews for colors used in cosmetics and the approval process may involve numerous studies to establish safety. FDA lists the approved colors in the Code of Federal Regulations (Title 21). These regulations describe the identity of the color, the allowed composition, the uses and restrictions and any other requirement necessary to ensure safe use.

Colors are also regulated in other countries. For instance, in the European Union, colors are identified on a list of allowed ingredients for coloring purposes. The colors are identified by a colour index number rather than a descriptive name such as Yellow 5 or Red 7. Many other countries follow the European Union model for color regulation.

Key Words & Phrases 重点词汇

color additives n. 颜色添加剂
foundation[faʊnˈdeɪʃn] n. 基础；地基；基金会；根据；创立
tint[tɪnt] n. 色彩；浅色 vt. 染（发）；给……着色
blush[blʌʃ] vi. 脸红；感到惭愧 n. 脸红；红色；羞愧 vt. 使成红色
hue[hjuː] n. 色彩；色度；色调；叫声
mint-flavored n. 薄荷味的
dye[daɪ] n. 染料；染色 v. 染；把……染上颜色；被染色
pigment[ˈpɪɡmənt] n. 色素；颜料 vt. 给……着色 vi. 呈现颜色
cough syrup n. 止咳糖浆；咳嗽糖浆
cereal[ˈsɪriəl] n. 谷类，谷物；谷类食品；谷类植物 adj. 谷类的；谷类制成的
contact lenses n. 隐形眼镜
eyeliner[ˈaɪlaɪnər] n. 眼线笔
decorative[ˈdekəreɪtɪv] adj. 装饰性的；装潢用的
visual[ˈvɪʒuəl] adj. 视觉的，视力的；栩栩如生的
impression[ɪmˈpreʃn] n. 印象；效果，影响；压痕，印记；感想
palette[ˈpælət] n. 调色板；颜料
drab[dræb] adj. 单调的；土褐色的 n. 浅褐色；无生气；邋遢；小额 vt. 使无生气
unstable[ʌnˈsteɪbl] adj. 不稳定的；动荡的；易变的
fade[feɪd] v. 褪色；凋谢；逐渐消失；使褪色 n. (电影、电视)淡出、淡入
perception[pərˈsepʃn] n. 认识能力；知觉，感觉；洞察力；看法；获取
absorption[əbˈzɔːrpʃn] n. 吸收；全神贯注，专心致志
reflection[rɪˈflekʃn] n. 反射；沉思；映象
visible light n. 可见光
wavelength[ˈweɪvleŋθ] n. 波长
light receptor cell n. 光感细胞
archaeologist[ˌɑːrkiˈɑːlədʒɪst] n. 考古学家
ancient[ˈeɪnʃənt] adj. 古代的；古老的，过时的；年老的 n. 古代人；老人
egyptian[iˈdʒɪpʃn] adj. 埃及的；埃及人的；埃及语的 n. 埃及人；古埃及语

colorant[ˈkʌlərənt]　n．着色剂
indigo[ˈɪndɪgoʊ]　n．靛蓝，靛蓝染料；靛蓝色；槐蓝属植物　adj．靛蓝色的
turmeric[ˈtɜːrmərɪk]　n．姜黄，姜黄根；姜黄根粉，郁金根粉
paprika[pəˈpriːkə]　n．辣椒粉，红辣椒
saffron[ˈsæfrən]　n．藏红花；橙黄色　adj．藏红花色的，橘黄色的
coloring intensity　n．颜色强度；彩色亮度
strict[strɪkt]　adj．严格的；绝对的；精确的；详细的
approval[əˈpruːvl]　n．批准；认可；赞成
the Federal Food, Drug, and Cosmetic Act(FD & C Act)　美国食品药品化妆品法案
coal-tar hair dye　n．煤焦油染发剂
violation[ˌvaɪəˈleɪʃn]　n．违反；妨碍；侵害；违背；强奸
detain[dɪˈteɪn]　vt．拘留；留住；耽搁
address[əˈdres]　n．地址；网址；演讲，致辞；称呼；谈吐　v．写姓名地址；演说，向……致辞；处理(问题)；称呼；诉说；从事；忙于
specification[ˌspesɪfɪˈkeɪʃn]　n．规格；说明书；详述
restriction[rɪˈstrɪkʃn]　n．限制；约束；束缚
certification[ˌsɜːrtɪfɪˈkeɪʃn]　n．证明，鉴定；出具课程结业证书，颁发证书
batch[bætʃ]　n．一批；一炉；一次所制之量　vt．分批处理
identity[aɪˈdentəti]　n．身份；同一性，一致；特性；恒等式
the code of federal regulations(CFR)　n．联邦政府管理条例
pertain[pərˈteɪn]　v．适合；关于；适用；从属，归属；生效，存在
maximum permissible concentration　n．最大允许浓度
the Office of Cosmetics and Colors　n．化妆品和色素办公室
FDA's Center for Food Safety and Applied Nutrition　n．食品药品管理局食品安全和应用营养中心
conduct[kənˈdʌkt]　v．组织，实施，进行；指挥(音乐)；带领，引导；举止，表现；传导(热或电)　n．行为举止；管理(方式)，实施(办法)；引导
review[rɪˈvjuː]　n．回顾；复习；评论；检讨；检阅　vt．回顾；检查；复审　vi．回顾；复习功课；写评论
composition[ˌkɑːmpəˈzɪʃn]　n．作文，作曲；[材料]构成；合成物

Practical Activities　实践活动

Part A　Reading comprehension.

1. According to the passage, which of the following can be added or applied to color additives? (　　)
A. food　　　　　B. drug　　　　　C. cosmetic　　　　　D. all of the above
2. According to the passage, which of the following is the feature of modern color additives? (　　)
A. wide color palette　B. safely　　　　　C. stable　　　　　D. all of the above
3. According to the passage, how can visual perception of color occur? (　　)

A. products absorp and/or reflect of visible light
B. color interacts with the light receptor cells in the eye
C. light receptor cells send a message to the brain
D. all of the above
4. According to the passage, man-made colors have (   ) advantage over natural colors.
A. more stable   B. greater coloring intensity
C. without using plants harvested in the wild   D. all of the above.
5. According to the passage, which statement in the following is correct? (   )
A. All color additives used in cosmetics must be approved by FDA
B. For all color additives, every batch made must be certified by FDA if they are to be used in cosmetics marketed in US
C. Color additives can be used freely as long as they get the Approval
D. Color additives may be used only for the intended uses stated in the regulations that pertain to them
6. For color additives used in cosmetics, there must be a regulation specifically addressing (   ).
A. intended uses   B. specifications   C. restrictions   D. all of the above
7. The Code of Federal Regulations (Title 21) describes (   ).
A. the identity of the color   B. the allowed composition
C. the uses and restrictions   D. all of the above
8. According to the passage, which statement in the following is correct? (   )
A. Color additives are very safe when used properly
B. Approved color additives listed by FDA can be used safely at any concentration
C. Color additives are regulated similar all over the world
D. Color additives marked in the European Union must follow FDA's regulation

Part B Fill in the terms of color additive.

1. _____ are dyes, pigments, or other substances that can impart color when added or applied to a food, drug or cosmetic. They may be used to create a product image or recognition, the "mood" of the product, or other visual impressions.
2. In the European Union, colors are identified by a _____ number rather than a descriptive name.

Part C Translation exercise.

1. Color additives may be used to create a product image or recognition, the "mood" of the product, or other visual impressions.
2. These colors, made in the laboratory, were found to be much more stable with greater coloring intensity, meaning that less colorant could be used in the product to accomplish the same effect.
3. In addition to approval, for a number of color additives, every batch made must be certified by FDA if they are to be used in cosmetics (or any other FDA-regulated product) marketed in the U.S.

4. Color additives are very safe when used properly.

5. These regulations describe the identity of the color, the allowed composition, the uses and restrictions and any other requirement necessary to ensure safe use.

Dyes and Pigments

Task 2 Color Additive Categories

任务二　颜色添加剂目录

Color additives are subject to a strict system of approval under U. S. law [Federal Food, Drug, and Cosmetic Act (FD & C Act), sec. 721; 21 U. S. C. 379e]. All color additives used in cosmetics (or any other FDA-regulated product) must be approved by FDA. In addition to approval, a number of color additives must be batch certified by FDA if they are to be used in cosmetics (or any other FDA-regulated product) marketed in US All color additives must meet the requirements for identity and specifications stated in the Code of Federal Regulations (CFR). Color additives may be used only for the intended uses stated in the regulations that pertain to them. The regulations also specify other restrictions for certain colors, such as the maximum permissible concentration in the finished product.

The FD & C Act Section 721(c)[21 U. S. C. 379e(c)] and color additive regulations [21 CFR Parts 70 and 80] separately approved color additives into two main categories: those subject to certification (sometimes called "certifiable") and those exempt from certification. In addition, the regulations refer to other classifications, such as straight colors and lakes.

Colors subject to certification. These color additives are derived primarily from petroleum and are sometimes known as "coal-tar dyes" or "synthetic-organic" colors. (Note: coal-tar colors are materials consisting of one or more substances that either are made from coal-tar or can be derived from intermediates of the same identity as coal-tar intermediates.) These certified colors generally have three-part names. The names include a prefix FD & C, D & C, or external D & C; a color, and a number. An example is "FD & C Yellow No. 5". Certified colors also may be identified in cosmetic ingredient declarations by color and number alone, without a prefix (such as "Yellow 5").

Colors exempt from certification. These color additives are obtained primarily from mineral, plant, or animal sources. They are not subject to batch certification requirements. However, they still are considered artificial colors, and when used in cosmetics or other FDA-regulated products, they must comply with the identity, specifications, uses, restrictions, and labeling requirements stated in the regulations.

Straight color. "Straight color" refers to any color additive listed in 21 CFR 73, 74, and 81. Here, 21 CFR 73, 74, and 81 [21 CFR 70.3(j)] means Title 21 of the Code of Federal Regulations Parts 73,74,81 and 82.

Lake. A lake is a straight color extended on a substratum by adsorption, coprecipitation, or chemical combination that does not include any combination of ingredients made by a simple mixing process [21 CFR 70.3(1)]. Because lakes are not soluble in water, they often are used when it is important to keep color from "bleeding", as in lipstick. In some cases, special restrictions apply to their use. As with any color additive, it is important to check the Summary of Color Additives Listed for Use in the United States in Foods, Drugs, Cosmetics and Medical Devices and the regulations themselves [21 CFR 82, Subparts B and C] to be sure you are using lakes only for their approved uses.

Key Words & Phrases　重点词汇

batch[ætʃ]　n. 一批；一炉；一次所制之量　vt. 分批处理
certify['sɜːrtɪfaɪ]　v. 证明；保证
separate['seprət]　adj. 分开的；单独的；不同的；各自的；不受影响的　v.（使）分离，分开；隔开；分手；（使）分居；（使）区别（于）　n. 可搭配穿着的单件衣服；抽印本；独立音响设备；土壤划分
category['kætəɡɔːri]　n. 种类，分类；[数学]范畴
subject to 服从；受制于
certifiable['sɜːrtɪfaɪəbl]　adj. 可证明的；可确认的；可保证的
exempt[ɪɡ'zempt]　adj. 被免除（责任或义务）的，获豁免的　v. 免除，豁免　n. 被免除义务者（尤指被免税者）
colors exempt from certification　n. 免检颜色
straight color　n. 纯品色
lake[leɪk]　n. 湖；深红色颜料；胭脂红；色淀　v.（使）血球溶解
petroleum[pə'troʊliəm]　n. 石油
coal-tar dyes　n. 煤焦油染料
synthetic-organic color　n. 有机合成的颜色
intermediate[ˌɪntər'miːdiət]　n. 中间，中级；中间体，中间片
prefix['priːfɪks]　n. 前缀　vt. 加前缀；将某事物加在前面
external[ɪk'stɜːrnl]　adj. 外部的；表面的；[药]外用的；外国的；外面的　n. 外部；外观；外面
declaration[ˌdeklə'reɪʃn]　n.（纳税品等的）申报；宣布；公告；申诉书
mineral['mɪnərəl]　n. 矿物；（英）矿泉水；无机物　adj. 矿物的；矿质的

artificial[ˌɑːrtɪˈfɪʃl]　*adj*．人造的；仿造的；虚伪的；非原产地的；武断的

extend on　　延长；延期；扩大；伸展

substratum[ˈsʌbstreɪtəm]　*n*．基础；根据；下层

coprecipitation[ˌkəupriˌsipiˈteiʃən]　*n*．共沉淀

chemical combination　　化合；化合作用

mix[mɪks]　*v*．（使）混和；配制；参与；交往，交际；混合录音；混成　*n*．混合；良莠不齐；混合物

bleed[bliːd]　*vt*．使出血；榨取　*vi*．流血；渗出；悲痛

lipstick[ˈlɪpstɪk]　*n*．口红；唇膏　*vt*．涂口红　*vi*．涂口红

Practical Activities　实践活动

Part A　Reading comprehension.

1. According to the passage, FDA separate approved color additives into two main categories：(　　) and (　　).

A. colors subject to certification　　B. colors exempt from certification

C. straight color　　D. lake

2. According to the passage, which following statement of the colors subject to certification is incorrect? (　　)

A. They are primarily derived from petroleum

B. They are synthetic-organic colors

C. They generally have three-part names

D. β-Carotene is a color subject to certification

3. According to the passage, which following statement of the colors exempt from certification is correct? (　　)

A. They are primarily derived from petroleum

B. They are primarily from mineral, plant, or animal sources

C. They are subject to batch certification requirements

D. They are considered natural colors

4. According to the passage, which following statement of the colors exempt from certification is incorrect? (　　)

A. They are considered artificial colors

B. They are not subject to batch certification requirements

C. Yellow 5 is a color exempt from certification

D. β-Carotene is a color exempt from certification

5. According to the passage, which statement in the following of straight color is correct? (　　)

A. All colors subject to certifications are straight colors

B. All colors exempt from certification are straight colors

C. All straight colors are colors subject to certification

D. All straight colors are colors exempt from certification

6. According to the passage, a lake can't be obtained by (   ).
A. a straight color extended on a substratum by adsorption
B. a straight color extended on a substratum by coprecipitation
C. a straight color extended on a substratum by chemical combination
D. a straight color simple mix with a substratum
7. According to the passage, a lake is (   ) in water.
A. soluble   B. insoluble   C. bleeding   D. mixing
8. According to the passage, (   ) is a color subject to certification.
A. annatto   B. iron oxides
C. D & C Black No. 2   D. silver
9. According to the passage, (   ) is a color exempt from certification.
A. FD & C Red No. 4   B. Ext. D & C Yellow No. 7
C. D & C Black No. 2   D. titanium dioxide
10. According to the passage, (   ) is not a straight color.
A. zinc oxide   B. mica
C. D & C Black No. 2   D. FD & C Blue No. 1

Part B Translation exercise.
1. These certified colors generally have three-part names. The names include a prefix FD & C, D & C, or External D & C; a color; and a number.
2. Colors exempt from certification are not subject to batch certification requirements. However, they still are considered artificial colors, and when used in cosmetics or other FDA-regulated products, they must comply with the identity, specifications, uses, restrictions, and labeling requirements stated in the regulations.
3. A lake is a straight color extended on a substratum by adsorption, coprecipitation, or chemical combination that does not include any combination of ingredients made by a simple mixing process.

知识拓展

Titanium Dioxide

Lesson 12 Auxiliary Materials
辅助性原料

Lead in 课前导入

Thinking and talking:
1. What are auxiliary materials? What are their roles in cosmetic products?
2. Is parabens and fragrance unsafe? Why do we have to put them in cosmetics?
3. What's the requirement of water for preparing cosmetic products? Why?

Related words:

auxiliary materials	辅助性原料	preservative	防腐剂
parabens	尼泊金酯类	microbial contamination	微生物污染
solvent	溶剂	water/aqua	水
scent	气味	fragrances	香料
consistency	黏度	carbomer	卡波姆

Task 1 Parabens in Cosmetics
任务一 化妆品中的防腐剂尼泊金酯类

A preservative is a natural or synthetic ingredient that is added to products such as foods, pharmaceuticals and personal care products to prevent spoilage, whether from microbial growth or undesirable chemical changes. The use of preservatives is essential in most products to prevent product damage caused by microorganisms and to protect the product from inadvertent contamination by the consumer during use. An ingredient that protects the product from the growth of microorganisms is called an antimicrobial. Without preservatives, cosmetic products, just like food, can become contaminated, leading to product spoilage and possibly irritation or infections. Microbial contamination of products, especially those used around the eyes and on the skin, can cause significant problems. Preservatives help prevent such problems.

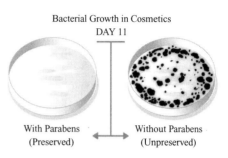

Bacterial Growth in Cosmetics
DAY 11
With Parabens (Preserved) ↔ Without Parabens (Unpreserved)

Ⅰ. Overview

Parabens are a group of commonly used ingredients that acts as a preservative in cosmetics and personal care products. They are highly effective in preventing the growth of fungi, bacteria, and yeast that could cause products to spoil, thus enhancing the shelf life and safety of products, and have been used safely for decades. Any product that contains water is susceptible to being spoiled by the growth of fungi or bacteria, which could cause problems such as discoloration, malodor, or breakdown of the product. Under certain conditions, an inadequately preserved product can become contaminated, allowing harmful microorganisms to grow. Parabens are widely used in all types of cosmetics to prevent these changes and protect the families who enjoy these products.

The parabens used most commonly in cosmetics are methylparaben, propylparaben, butylparaben, and ethylparaben. Product ingredient labels typically list more than one paraben in a product, and parabens are often used in combination with other types of preservatives to better protect against a broad range of microorganisms.

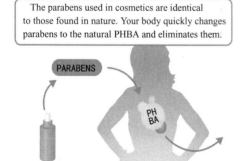

The parabens used in cosmetics are identical to those found in nature. Your body quickly changes parabens to the natural PHBA and eliminates them.

Parabens were first introduced in 1923 and are used as preservatives in the food, drug, and personal care industries. They all share para-hydroxybenzoic acid (PHBA) as a common chemical structure. PHBA occurs naturally in many fruits and vegetables, such as cucumbers, cherries, carrots, blueberries, and onions, and is also formed naturally in your body by the breakdown of certain amino acids. The parabens used in cosmetics are identical to those found in nature, and are quickly eliminated by the body.

According to the 2019 U.S. Food and Drug Administration (FDA) Voluntary Cosmetic Registration Program (VCRP) data, methylparaben was used in 11,739 formulations (9,347 of which are leave-on formulations). Propylparaben had the next highest number of reported uses at 9,034 (7,520 of which are leave-on formulations).

II. Safety

Some have speculated whether there is a connection between parabens and cancer, suggesting parabens can act like estrogen, a common hormone, through a process called endocrine disruption. Scientific studies have shown this interaction to be very weak, observed only with extremely high doses, far greater than anyone would be exposed to under actual conditions of use or with repeated use. Many materials found in plants used as food also have a weak estrogenic effect in cellular studies. These naturally occurring materials are present in soy and other fruits and vegetables, and when tested in the same way as parabens, give similar results. In fact, some phytoestrogens have been shown to be 10,000 times more potent than parabens. Most scientists agree there is no endocrine disrupting effect from the use of parabens in cosmetics and personal care products because their action, if any, is so weak.

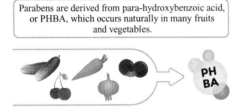

Parabens are derived from para-hydroxybenzoic acid, or PHBA, which occurs naturally in many fruits and vegetables.

Another myth is that parabens are banned outside the U.S. when, in fact, safe levels of parabens have been established and these ingredients are approved by government agencies-including in the European Union (EU), Japan, Australia, and Canada-for use in things like cosmetics and personal care products. The U.S. FDA has stated there is no reason for consumers to be concerned about the use of cosmetics containing parabens. The FDA has also classified methylparaben and propylparaben as GRAS, which means they are Generally Regarded As Safe for use in preserving food.

The FDA considers methylparaben and propylparaben to be Generally Recognized As Safe (GRAS) as antimicrobial agents in food [21CFR184.1490; 21CFR184.1670]. Butylparaben, ethylparaben, and propylparaben are approved for use in food for humans as synthetic flavoring substances and adjuvants. [21CFR172.515] Ethylparaben may be used as an indirect food additive as a component of adhesives and coatings. [21CFR175.105] In pharmaceutical drugs, methylparaben, ethylparaben, propylparaben, and butylparaben have been approved as inactive ingredients (excipients). An FDA ingredient webpage titled "Parabens in Cosmetics" concludes: "At this time, we do not have information showing that parabens, as they are used in cosmetics, have an effect on human health."

According to the general provisions of the Cosmetics Regulation of the European Union, the following paraben ingredients are allowed to be used as cosmetic preservatives at concentrations up to 0.4 percent (as acid) for single esters or 0.8 percent (as acid) for mixtures of esters: 4-hydroxybenzoic acid, methylparaben, potassium ethylparaben, potassium paraben, sodium methylparaben, sodium ethylparaben, ethylparaben, sodium paraben, potassium methylparaben, and calcium paraben. Propylparaben and butylparaben have maximum concentration limits of 0.19 percent (single esters and their salts). Isobutylparaben and its salts, isopropylparaben and its salts, phenylparaben, benzylparaben, and pentylparaben are prohibited from use in cosmetics products in the EU.

Key Words & Phrases 重点词汇

parabens[pəˈræbenz] n. 对羟苯甲酸酯，对羟基苯甲酸酯类，尼泊金类
preservative[prɪˈzɜːrvətɪv] n. 防腐剂；预防法；防护层 adj. 防腐的；有保存力的；有保护性的
pharmaceutical[ˌfɑːrməˈsuːtɪkl] adj. 制药(学)的 n. 药物
spoilage[ˈspɔɪlɪdʒ] n. 损坏，糟蹋；掠夺；损坏物
microbial[maɪˈkroʊbɪəl] adj. 微生物的；由细菌引起的
essential[ɪˈsenʃl] adj. 基本的；必要的；本质的；精华的 n. 本质；要素；要点；必需品
microorganism[ˌmaɪkroʊˈɔːrgənɪzəm] n. 微生物；微小动植物
inadvertent[ˌɪnədˈvɜːrtnt] adj. 疏忽的；不注意的；无意中做的
contamination[kənˌtæmɪˈneɪʃn] n. 污染，玷污；污染物
antimicrobial[ˌæntimaɪˈkroʊbɪəl] adj. 杀菌的，抗菌的 n. 杀菌剂，抗菌剂
infection[ɪnˈfekʃn] n. 感染；传染；影响；传染病
microbial contamination 微生物污染
fungi[ˈfʌŋgi] n. 真菌；菌类；蘑菇(fungus 的复数)
bacteria[bækˈtɪrɪə] n. 细菌
yeast[jiːst] n. 酵母；泡沫；酵母片；引起骚动因素
spoil[spɔɪl] vt. 溺爱；糟蹋；掠夺 vi. 掠夺；变坏；腐败 n. 次品；奖品
shelf life n. 保存限期；[电学]贮藏寿命；闲置时间；适用期
susceptible[səˈseptəbl] adj. 易受影响的；易感动的；容许……的 n. 易得病的人
discoloration[dɪsˌkʌləˈreɪʃn] n. 变色；污点
malodor[mælˈodə] n. 恶臭；臭气
methylparaben[ˌmeθɪlpəˈræben] n. 对羟基苯甲酸甲酯
propylparaben[ˌprəʊpɪlˈpɑːrəben] n. 尼泊金丙酯；对羟基苯甲酸丙酯
butylparaben[ˌbetɪlpəˈræben] n. 尼泊金丁酯；对羟基苯甲酸丁酯
ethylparaben[ˌeθɪlˈpærəben] n. 对羟基苯甲酸乙酯
para-hydroxybenzoic acid(phba) n. 对羟基苯甲酸(phba)
amino acid[əˈmiːnəʊ ˈæsɪd] n. 氨基酸
voluntary cosmetic registration program(vcrp)
estrogen[ˈɛstrədʒən] n. 雌性激素
hormone[ˈhɔːrmoʊn] n. 激素，荷尔蒙
endocrine disruption 内分泌干扰
potent[ˈpoʊtnt] adj. 有效的；强有力的，有权势的；有说服力的
ban[bæn] v. 禁止，取缔；(官方)把(某人)逐出某地 n. 禁止，禁令，禁忌；剥夺公民权的判决；诅咒；巴尼(罗马尼亚货币单位)

synthetic flavoring substance　合成调味物质
adjuvant['ædʒəvənt]　adj. 辅助的　n. 佐药；辅助物
excipient[ɪk'sɪpɪənt]　n. [药]赋形剂
potassium ethylparaben　羟苯乙酯钾
potassium paraben　对羟基苯甲酸钾
sodium methylparaben　羟苯甲酸钠甲酯
sodium ethylparaben　羟苯乙酯钠
sodium paraben　对羟基苯甲酸钠
potassium methylparaben　对羟基苯甲酸甲酯钾
calcium paraben　对羟基苯甲酸钙
isobutyl paraben　对羟基苯甲酸异丁酯
isopropyl paraben　对羟基苯甲酸异丙酯
phenyl paraben　对苯二甲酸苯酯
benzyl paraben　对羟基苯甲酸苄酯
pentyl paraben　对羟基苯甲酸戊酯

Practical Activities　实践活动

Part A　Reading comprehension.

1. According to the passage, preservatives can help prevent spoilage, which means they can (　　).

    A. prevent product damage caused by microorganism

    B. protect product from inadvertent contamination during use

    C. protect product from contamination

    D. all of the above

2. According to the passage, microbial contamination of products possibly causes (　　).

    A. irritation　　　　　　　　　　B. infections

    C. all of the above　　　　　　　D. none of the above

3. According to the passage, fungi, bacteria, and yeast are (　　).

    A. parabens　　　　　　　　　　B. preservatives

    C. microorganisms　　　　　　　D. personal care products

4. According to the passage, symptoms of spoilage including (　　).

    A. discoloration　　　　　　　　B. malodor

    C. breakdown of the product　　　D. all of the above.

5. According to the passage, methylparaben often used in combination with propylparaben to (　　).

    A. save cost

    B. protect against a broad range of microorganisms

    C. share a common chemical structure

D. none of the above

6. According to the passage, which statement in the following is correct? (   )
A. PHBA occurs only in parabens
B. PHBA occurs in many fruits and vegetables
C. PHBA can't be eliminated by human body
D. PHBA is very harmful for human body

7. According to the passage, which statement in the following is incorrect? (   )
A. Endocrine disruption of parabens is very weak under actual conditions of use
B. Many materials found in plants used as food also have a weak estrogenic effect
C. Parabens are banned outside the US
D. Methylparaben and propylparaben are Generally Regarded As Safe for use in preserving food

8. According to the passage, methylparaben are allowed to be used as cosmetic preservatives at concentrations up to (   ) percent (as acid).
A. 0.4   B. 0.8   C. 0.19   D. none of the above

Part B   Fill in the blanks.

1. A _____ is a natural or synthetic ingredient that is added to products such as foods, pharmaceuticals and personal care products to prevent spoilage.

2. An ingredient that protects the product from the growth of microorganisms is called an _____ .

3. Without preservatives, cosmetic products can become contaminated, leading to product _____ and possibly _____ or _____ .

Part C   Translation exercise.

1. The use of preservatives is essential in most products to prevent product damage caused by microorganisms and to protect the product from inadvertent contamination by the consumer during use.

2. They are highly effective in preventing the growth of fungi, bacteria, and yeast that could cause products to spoil, thus enhancing the shelf life and safety of products.

3. The parabens used most commonly in cosmetics are methylparaben, propylparaben, butylparaben, and ethylparaben.

4. Product ingredient labels typically list more than one paraben in a product, and parabens are often used in combination with other types of preservatives to better protect against a broad range of microorganisms.

5. At this time, we do not have information showing that parabens as they are used in cosmetics have an effect on human health.

知识拓展

Allergens in Cosmetics

Task 2　Other Auxiliary Materials

任务二　认识其他常见辅助性原料

Ⅰ. Water

Water is used in the formulation of virtually every type of cosmetic and personal care product. It can be found in lotions, creams, bath products, cleansing products, deodorants, makeup, moisturizers, oral hygiene products, personal cleanliness products, skin care products, shampoo, hair conditioners, shaving products, and suntan products.

Water is primarily used as a solvent in cosmetics and personal care products in which it dissolves many of the ingredients that impart skin benefits, such as conditioning agents and cleansing agents. Water also forms emulsions in which the oil and water components of the product are combined to form creams and lotions. These are sometimes referred to as oil-in-water emulsions or as water-in-oil depending on the ratios of the oil phase and water phase. Only water that is free of toxins, pollutants and microbes is used in the formulation of cosmetics and personal care products. Water used for this purpose is also referred to as distilled water, purified water and aqua, the name used in the European Union.

Ⅱ. Fragrance

1. Overview

Fragrances are complex combinations of natural and/or man-made substances that are added to many consumer products to give them a distinctive smell. Fragrances are used in a wide variety of products to impart a pleasant odor, mask the inherent smell of some ingredients, and enhance the experience of using the product. Fragrances create important benefits that are ubiquitous, tangible, and valued. They solve important functional problems and they satisfy valued emotional needs. Fragrances can communicate complex ideas-creating mood, signalling cleanliness, freshness, or softness, alleviating stress, creating well-being, and triggering allure and attraction.

Fragrances have been enjoyed for thousands of years and contribute to people's individuality, self-esteem and personal hygiene. Consumer research indicates that fragrance is one of the key factors that affect people's preferences for cosmetic and personal care products. There are hundreds of fragrances created every year, in countries all over the world. Our sense of smell is directly connected to the brain's limbic system where our sense of memory and our emotions are stored. Numerous studies confirm that fragrances enhance well being and have a positive impact on the psyche. Often a particular fragrance becomes strongly associated with product identity and acceptability.

Fragrance and flavor formulas are complex mixtures of many different natural and man-made chemical ingredients, and they are the kinds of cosmetic components that are considered to be "trade secrets." Under U. S. regulations, fragrance ingredients can be listed collectively simply as "Fragrance". Similarly, in the European Union (EU), perfume mixtures are labeled collectively as "parfum," except for 26 specific perfume ingredients which are required to be listed individually by name.

2. Function

Fragrance ingredients are also commonly used in other products, such as shampoos, shower gels, shaving creams, and body lotions. Even some products labeled "unscented" may contain fragrance ingredients. This is because the manufacturer may add just enough fragrance to mask the unpleasant smell of other ingredients, without giving the product a noticeable scent.

3. Safety

Fragrance ingredients in cosmetics must meet the same requirement for safety as other cosmetic ingredients: they must be safe for consumers when they are used according to labeled directions, or as people customarily use them. This is a responsibility that fragrance manufacturers and the companies that use fragrances in their products take very seriously. Fragrance and cosmetic/personal care product manufacturers are diligent in assessing the safety of fragrances and their ingredients. A very small percentage of individuals may be allergic or sensitive to certain ingredients in cosmetics, food, or other products, even if those ingredients are safe for most people. Although some substances may have the potential to cause allergic reactions, they can still be formulated into consumer products at safe levels. This is the case for fragrance ingredients.

Some components of fragrance formulas may have a potential to cause allergic skin reactions (i. e., dermal sensitization) or sensitivities for some people. Based on the chemical, cellular and molecular understanding of dermal sensitization, it is possible to conduct a safety assessment using a methodology known as Quantitative Risk Assessment (QRA) to determine safe use levels of fragrance ingredients in a variety of consumer product types. More information on QRA can be found on both the IFRA and RIFM websites.

Ⅲ. Carbomer

1. Overview

Carbomer is a term used for a series of polymers primarily made from acrylic acid. The carbomers are white, fluffy powders but are frequently used as gels in cosmetics and personal care products. Carbomers can be found in a wide variety of product types including skin, hair, nail, and makeup products, as well as dentifrices. The carbomers help to distribute or suspend an insoluble solid in a liquid. They are also used to keep emulsions from separating into their oil and liquid components. Carbomers are often used to control the consistency and flow of cosmetics and personal care products.

The carbomers are all chemically similar, differing from each other in molecular weight and viscosity. Carbomers have the ability to absorb and retain water, and these polymers can swell to many times their original volume. On a label of a cosmetic or personal care product, the name carbomer may be associated with a number such as 910,934,940,941 and 934P. These numbers are an indication of molecular weight and the specific components of the polymer.

2. Safety

The safety of the carbomer has been assessed by the Cosmetic Ingredient Review (CIR) Expert Panel. The CIR Expert Panel evaluated the scientific data and concluded that carbomer polymers were safe as ingredients in cosmetics and personal care products.

CIR Safety Review: The CIR Expert Panel reviewed acute oral studies showing that carbomer polymers have low toxicities when ingested. Minimal skin irritation and no to moderate eye irritation were observed. Subchronic feeding studies with a carbomer polymer resulted in lower than normal body weights, but no abnormal changes were observed in the organs. Some gastrointestinal irritation and marked pigment deposition within specific cells in the liver, called Kupffer cells, were seen in studies with carbomer. Clinical studies with carbomers showed that these polymers have low potential for skin irritation and sensitization at concentrations up to 100%. A carbomer polymer demonstrated low potential for phototoxicity and photo-contact allergenicity.

Key Words & Phrases 重点词汇

Ⅰ. Water

virtually[ˈvɜːrtʃuəli] adv. 事实上，几乎；实质上
oral hygiene product 口腔卫生产品
shaving product 剃须产品
suntan product 防晒产品
impart[ɪmˈpɑːrt] vt. 给予(尤指抽象事物)，传授；告知，透露
phase[feɪz] n. 月相；时期，阶段 vt. 分阶段进行；使定相
toxin[ˈtɑːksɪn] n. 毒素；毒质
pollutant[pəˈluːtənt] n. 污染物
microbe[ˈmaɪkroʊb] n. 细菌，微生物
distilled water 蒸馏水
purified water 净化水
aqua[ˈɑːkwə] n. 水；溶液；浅绿色 adj. 浅绿色的

Ⅱ. Fragrance

fragrance[ˈfreɪɡrəns] n. 香味，芬芳
distinctive[dɪˈstɪŋktɪv] adj. 独特的，有特色的；与众不同的
mask[mæsk] n. 面具；口罩；掩饰 vi. 掩饰；戴面具；化装 vt. 掩饰；戴面具；使模糊
inherent[ɪnˈherənt] adj. 固有的；内在的；与生俱来的，遗传的
ubiquitous[juːˈbɪkwɪtəs] adj. 普遍存在的；无所不在的

tangible[juːˈbɪkwɪtəs]　*adj*．普遍存在的；无所不在的
emotional[ɪˈmoʊʃənl]　*adj*．情绪的；易激动的；感动人的
signal[ˈsɪɡnəl]　*n*．信号；暗号；导火线　*vt*．标志；用信号通知　*adj*．显著的；作为信号的　*vi*．发信号
cleanliness[ˈklenlɪnəs]　*n*．清洁
freshness[ˈfreʃnəs]　*n*．新；新鲜；精神饱满
softness[ˈsɔːftnəs]　*n*．温柔；柔和
alleviat[əˈliːvieɪt]　*vt*．减轻，缓和
well-being[ˈwel biːɪŋ]　*n*．幸福；康乐
trigger[ˈtrɪɡər]　*n*．扳机；起因，引起反应的事；触发器，引爆装置　*v*．触发，引起；开动(装置)
allure[əˈlʊr]　*n*．诱惑力；引诱力；吸引力　*v*．吸引；引诱
attraction[əˈtrækʃn]　*n*．吸引，吸引力；引力；吸引人的事物
individuality[ˌɪndɪˌvɪdʒuˈæləti]　*n*．个性；个人；个人特征；个人的嗜好(通常复数)
self-esteem[ˌself ɪˈstiːm]　*n*．自尊；自负；自大
hygiene[ˈhaɪdʒiːn]　*n*．卫生；卫生学；保健法
limbic[ˈlɪmbɪk]　*adj*．边的；缘的
emotion[ɪˈmoʊʃn]　*n*．情感；情绪
psyche[ˈsaɪki]　*n*．灵魂；心智
flavor[ˈfleɪvər]　*n*．情味，风味；香料；滋味　*vt*．加味于
trade secret　*n*．商业秘密；行业秘密
unscented[ʌnˈsentɪd]　*adj*．无香味的；无气味的
unpleasant[ʌnˈpleznt]　*adj*．讨厌的；使人不愉快的
smell[smel]　*n*．气味，嗅觉；臭味　*v*．嗅，闻；有……气味；察觉到；发出……的气味
noticeable[ˈnoʊtɪsəbl]　*adj*．显而易见的，显著的；值得注意的
diligent[ˈdɪlɪdʒənt]　*adj*．勤勉的；用功的，费尽心血的
cellular[ˈseljələr]　*adj*．细胞的；多孔的；由细胞组成的　*n*．移动电话；单元
methodology[ˌmeθəˈdɑːlədʒi]　*n*．方法学，方法论
quantitative risk assessment(qra)　定量风险评价

Ⅲ．Carbomer

acrylic acid　[有机化学]丙烯酸
fluffy[ˈflʌfi]　*adj*．蓬松的；松软的；毛茸茸的；无内容的
dentifrice[ˈdentəfrɪs]　*n*．牙膏；牙粉
molecular weight　分子量
viscosity[vɪˈskɑːsəti]　*n*．黏性；黏度
retain[rɪˈteɪn]　*vt*．保持；雇；记住
volume[ˈvɑːljuːm]　*n*．量；体积；卷；音量；大量；册　*adj*．大量的　*vi*．成团卷起　*vt*．把……收集成卷
ingest[ɪnˈdʒest]　*vt*．摄取；咽下；吸收；接待

subchronic[sʌbˈkrɒnɪk] adj.[医]亚慢性的
abnormal[æbˈnɔːrml] adj. 反常的，不规则的；变态的
organ[ˈɔːrɡən] n.[生物]器官；机构；风琴；管风琴；嗓音

Practical Activities 实践活动

Part A Reading comprehension.

1. Water is primarily used as a (   ) in cosmetics and personal care products.
A. solvent　　　　　　　　　　　B. conditioning agents
C. cleansing agents　　　　　　　D. none of the above

2. According to the passage, oil-in-water emulsions or as water-in-oil depending on (   ).
A. the ratios of water phase
B. the ratios of oil phase
C. the ratios of the oil phase and water phase
D. none of the above

3. According to the passage, water used in the formulation of cosmetics and personal care products should be (   ).
A. distilled water　　　　　　　　B. purified water
C. aqua　　　　　　　　　　　　D. all of the above

4. According to the passage, aqua used in the formulation of cosmetics and personal care products can contain (   )
A. water　　　B. toxins　　　C. pollutants　　　D. microbes

5. According to the passage, fragrances are used to (   ).
A. impart a pleasant odor
B. mask the inherent smell of some ingredients
C. enhance the experience of using the product
D. all of the above

6. According to the passage, which statement in the following is incorrect? (   )
A. Fragrance satisfy valued emotional needs
B. Fragrance is one of the key factors that affect people's preference for cosmetic and personal care products
C. Fragrances enhance well being and have a positive impact on the psyche
D. Fragrances associated little with product identity and acceptability

7. According to the passage, which statement in the following is correct? (   )
A. Products labeled "unscented" don't contain fragrance ingredients
B. Fragrances always give products a noticeable scent
C. It's the government's responsibility to take the safety of fragrances
D. If some substances have the potential to cause allergic reactions, they can't be formulated into consumer products

8. According to the passage, which statement in the following is incorrect? (   )
A. Carbomers are derived from acrylic acid

Lesson 12　Auxiliary Materials

B. Carbomer are frequently used as white, fluffy powder in cosmetics and personal care products

C. Carbomers help to distribute or suspend an insoluble solid in a liquid

D. Carbomers are often used to control the consistency and flow of cosmetics and personal care products

9. According to the passage, the number such as 910, 934, 940, 941 and 934P associated with carbomer is an indication of (　　).

A. molecular weight　　　　　　B. specific components
C. all of the above　　　　　　　D. none of the above

Part B　Fill in the blanks.

1. _____ are complex combinations of natural and/or man-made substances that are added to many consumer products to give them a distinctive smell.

2. Under U.S. regulations, fragrance ingredients can be listed collectively simply as "_____". Similarly, in the European Union (EU), perfume mixtures are labeled collectively as "_____".

3. Carbomers are often used to control the _____ and _____ of cosmetics and personal care products.

Part C　Translation exercise.

1. Water is primarily used as a solvent in cosmetics and personal care products.

2. Water also forms emulsions in which the oil and water components of the product are combined to form creams and lotions. These are sometimes referred to as oil-in-water emulsions or as water-in-oil depending on the ratios of the oil phase and water phase.

3. Fragrances are used in a wide variety of products to impart a pleasant odor, mask the inherent smell of some ingredients, and enhance the experience of using the product.

4. The Carbomers help to distribute or suspend an insoluble solid in a liquid.

5. Carbomers are often used to control the consistency and flow of cosmetics and personal care products.

6. Carbomers have the ability to absorb and retain water, and these polymers can swell to many times their original volume.

7. Carbomer polymers have low toxicities when ingested.

8. Minimal skin irritation and no to moderate eye irritation were observed.

9. Clinical studies with Carbomers showed that these polymers have low potential for skin irritation and sensitization at concentrations up to 100%.

10. A Carbomer polymer demonstrated low potential for phototoxicity and photo-contact allergenicity.

知识拓展

认识几种化妆品原料

Lesson 13 Plant Extracts
认识植物提取物

Lead in 课前导入

Thinking and talking:
1. How many types of plant extracts do you know?
2. What are their characteristics?

Related words:

plant extracts	植物提取物	multifunctional	多功能的
alpha hydroxy acids	果酸		
plant-derived (botanical) ingredients		植物成分	
naturally occurring products		天然产品	

Task 1 Alpha Hydroxy Acids

任务一 认识果酸

I. Overview

Alpha hydroxy acids (AHAs) are a class of chemical compounds that occur naturally in fruits, milk, and sugar cane. Although they are called acids they are not to be confused with strong industrial acids such as hydrochloric acid and sulfuric acid. The AHAs most commonly used in cosmetic products are glycolic acid (which is derived from sugar cane) and lactic acid (the substance that gives you muscle burn when you exercise). Other AHAs used include citric acid (from oranges, lemons, etc.), 2-hydroxyoctanoic acid, and 2-hydroxydecanoic acid. The AHAs may be obtained from their natural sources or may be made synthetically.

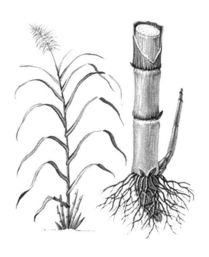

II. Function

Products containing AHA ingredients may be for consumer use, salon use, or medical use, depending on the concentration and pH (the degree of acidity or alkalinity). Products containing AHAs are marketed for a variety of purposes, such as smoothing fine lines and surface wrinkles, improving skin texture and tone, unblocking and cleansing pores, and improving skin condition in general. Since 1992 there have been products marketed as cosmetics intended to exfoliate and cleanse the skin. These products most often contain glycolic and lactic acids. They help reduce the appearance of skin wrinkling, even skin tones and soften and smoothe the skin. AHAs as used in cosmetics may function as exfoliants. They act on the surface of the skin by removing dead surface cells, thereby improving the appearance of the skin. In addition, lactic acid functions as a humectant-skin conditioning agent. AHAs also function as pH adjusters. pH Adjusters are materials added to products to make sure they are not too acid or base (low pH and high pH) and are thus mild and non-irritating. Many AHAs are naturally occurring products. For example, glycolic acid, a constituent of sugar cane juice, and lactic acid, which occurs in sour milk, molasses, apples and other fruits, tomato juice, beer, and wines, are carboxylic acid that function as pH adjusters and mild exfoliants in various types of cosmetic formulations.

III. Safety

AHAs have been safely used for many years (in fact the use of sour milk baths containing

lactic acid goes back to the ancient Egyptians). In response to concerns raised by FDA, the cosmetic industry asked the Cosmetic Ingredient Review (CIR) to conduct an independent assessment of the safety of AHAs. They recommended that cosmetics containing AHA ingredients be formulated so as to avoid increasing sun sensitivity or to provide directions for use that include the daily use of sun protection (a SPF product). After considering this request, FDA published formal guidance in July 2005, that encourages manufacturers to label their products with the following statement: "Sunburn Alert: The product contains an alpha hydroxy acid (AHA) that may increase your skin's sensitivity to the sun and particularly the possibility of sunburn. Use a sunscreen, wear protective clothing, and limit sun exposure while using this product and for a week afterwards."

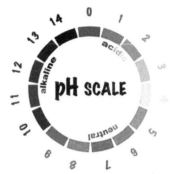

universal indicator pH color chart

The CIR Expert Panel concluded that AHA ingredients are safe for products formulated for home use when used at lower concentrations and controlled pH of the formula and for professional use at higher concentrations at slightly lower pH. The panel further required that home use products be formulated to avoid increased sun sensitivity or that the use of daily sun protection is advised. For professional products, the panel recommended that they are applied by a trained professional for brief and discontinuous use followed by directions for the daily use of sun protection.

Key Words & Phrases　重点词汇

alpha hydroxy acids　*n.* 果酸
sugar cane[ʃʊgɚ ken]　*n.* 甘蔗
hydrochloric acid[haɪdrəklɔrɪk ˈæsɪd]　*n.* 盐酸
confuse[kənˈfjuːz]　*vt.* 使混乱；使困惑
sulfuric acid[sʌlˈfjʊrɪk ˈæsɪd]　*n.* 硫酸
glycolic acid[glaɪˈkɑlɪk]　*n.* 羟基乙酸
lactic acid[ˈlæktɪkˈæsɪd]　*n.* 乳酸
muscle[ˈmʌsl]　*n.* 肌肉；力量　*vt.* 加强；使劲搬动；使劲挤出　*vi.* 使劲行进
citric acid[ˈlæktɪkˈæsɪd]　*n.* 柠檬酸
2-hydroxyoctanoic acid　2-羟基辛酸
2-hydroxydecanoic acid　2-羟基癸酸
salon[səˈlɑːn]　*n.* 沙龙；客厅；画廊；美术展览馆
acidity[əˈsɪdəti]　*n.* 酸度；酸性；酸过多；胃酸过多

Lesson 13　Plant Extracts

alkalinity[ˌælkəˈlɪnəti]　*n.*［化学］碱度；碱性
texture[ˈtekstʃər]　*n.* 质地；纹理；结构；本质，实质
tone[toʊn]　*n.* 语气；色调；音调；音色　*vt.* 增强；用某种调子说　*vi.* 颜色调和；呈现悦目色调
skin texture and tone　皮肤质地和色调
pore[pɔːr]　*n.*（皮肤上的）毛孔，黑头；（植物的）气孔，孔隙
exfoliate[eksˈfoʊlieɪt]　*vi.* 片状剥落；鳞片样脱皮　*vt.* 使片状脱落；使鳞片状脱落
dead surface cells　脱落坏死的表面细胞
pH adjusters　pH 值调节剂
sour[ˈsaʊər]　*adj.* 酸的；发酵的；刺耳的；酸臭的；讨厌的　*vi.* 发酵；变酸；厌烦
molasses[məˈlæsɪz]　*n.* 糖蜜，糖浆
carboxylic acid　*n.*［有机化学］羧酸
alert[əˈlɜːrt]　*adj.* 警惕的，警觉的；机警的，机敏的　*v.* 使警觉，警告；使意识到　*n.* 警戒，警惕；警报
sunscreen[ˈsʌnskriːn]　*n.*（防晒油中的）遮光剂；防晒霜
brief[briːf]　*adj.* 简短的，简洁的；短暂的，草率的　*n.* 摘要，简报；概要，诉书
discontinuous[ˌdɪskənˈtɪnjuəs]　*adj.* 不连续的；间断的

Practical Activities　实践活动

Part A　Reading comprehension.

1. According to the passage, alpha hydroxy acids (AHAs) are (　　).
 A. strong industrial acids　　　　B. only obtained from natural sources
 C. not made synthetically　　　　D. not strong acids
2. According to the passage, the most commonly used AHAs in cosmetic products are (　　) and (　　).
 A. citric acid　　　　B. glycolic acid
 C. lactic acid　　　　D. none of the above
3. According to the passage, glycolic acid and lactic acid are all (　　).
 A. bases　　　　B. carboxylic acids
 C. humectant-skin conditioning agents　　　　D. strong acids
4. According to the passage, which statement in the following is incorrect? (　　)
 A. AHAs can help avoid increasing sun sensitivity
 B. AHAs can increase skin's sensitivity to the sun
 C. AHAs can increase the possibility of sunburn
 D. Customers usde AHAs should limit sun exposure
5. According to the passage, which statement in the following is incorrect? (　　)
 A. AHAs formulated for home use should at lower concentrations
 B. AHAs formulated for professional use can at higher concentrations
 C. PH is higher for home use products containing AHAs
 D. PH is lower for home use products containing AHAs

Part B Fill in the blanks.

1. The degree of acidity or alkalinity is called _____ .
2. Products containing AHA ingredients may be for consumer use, salon use, or medical use, depending on the _____ and _____ .
3. _____ are materials added to products to make sure they are not too acid or base (low pH and high pH) and are thus mild and non-irritating.

Part C Translation exercise.

1. Products containing AHA ingredients may be for consumer use, salon use, or medical use, depending on the concentration and pH (the degree of acidity or alkalinity).
2. AHAs as used in cosmetics may function as exfoliants. They act on the surface of the skin by removing dead surface cells, thereby improving the appearance of the skin.
3. In addition, lactic acid functions as a humectant-skin conditioning agent.
4. AHAs also function as pH adjusters.
5. The product contains an alpha hydroxy acid (AHA) that may increase your skin's sensitivity to the sun and particularly the possibility of sunburn.
6. AHA ingredients are safe for products formulated for home use when used at lower concentrations and controlled pH of the formula and for professional use at higher concentrations at slightly lower pH.

Part D Summary.

What are the purposes of products containing AHAs? Try to summarize them.

Task 2 Other Plant Extracts

任务二 认识其他常见植物提取物

Ⅰ. Trehalose

Molecular weight: 342.30.

Melting point: The dihydrate melts at 97℃; the anhydrous melts at 210℃.

Solubility: Very soluble in water, formamide, and dimethyl

sulfoxide; soluble in hot alcohol.

Stability: Very stable and chemically unreactive; does not dissociate into two reducing monosaccharidic constituents unless exposed to extreme hydrolytic conditions or to the actions of trehalase.

Natural occurrence and methods of preparation: Found in fungi, bacteria, yeasts, and insects; isolated from the ergot of rye; isolated from yeast. Produced from starch using the enzymes maltooligosyl-trehalose synthase and maltooligosyl-trehalose trehalohydrolase.

Functions: Trehalose has a number of reported functions in cosmetics, with the most common use as a skin-conditioning agent. Other commonly reported functions are used as a humectant or as a flavoring agent.

II. Ceramides

Ceramide ingredients function primarily as hair conditioning agents and skin conditioning agents-miscellaneous in cosmetics. Naturally-occurring ceramides are normal constituents of the skin and are essential components of the epidermal permeability barrier.

Ceramides are among the lipids that make up sphingomyelin, which is a major component of the lipid bilayer that forms cell membranes of cells in the stratum corneum. Thus, ceramides are lipophilic and likely to be absorbed into the skin.

The oral LD_{50} was greater than 2000 mg/kg in rat studies of ceramide 2, ceramide AP, hydroxypalmitoyl sphinganine, and 2-oleamido-1,3-octadecanediol, and greater than 5000 mg/kg in rat studies of ceramide NP. In dermal studies, the LD_{50} was greater than 2000 mg/kg in rats exposed to ceramide NP, ceramide AP, or 2-oleamido-1,3-octadecanediol. The Cosmetic Ingredient Review (CIR) Expert Panel concluded that ceramides were safe in cosmetics in the present practices of use and concentration, noting that ingredients derived from bovine central nervous system tissues are not permitted for use in cosmetic products. Although some ceramides are plentiful in bovine central nervous system tissues (e.g., brain and spinal cord), the U.S. Food and Drug Administration (FDA) prohibits the use of ingredients derived from such tissues in cosmetic products because of the risk of transmitting infectious agents, such as bovine spongiform encephalitis (BSE) (21 CFR 700.27). Some ceramide ingredients may be derived from plant sources (e.g., those designated as phytosphingosines in the INCI names or definitions), which do not pose the risks associated with ingredients derived from the bovine central nervous system.

III. Simmondsia Chinensis (Jojoba) Seed Oil

Simmondsia chinensis (jojoba) seed oil and simmondsia chinensis (jojoba) seed wax, also called jojoba oil and jojoba wax, are natural ingredients derived from the seeds of the desert shrub, Simmondsia chinensis. Other ingredients made from jojoba oil include simmondsia chinensis (jojoba) butter, hydrogenated jojoba oil, jojoba esters, hydrolyzed jojoba esters, isomerized jojoba oil and jojoba alcohol. synthetic jojoba oil is amixture of wax esters of fatty acids and alcohols that is indistinguishable from natural jojoba oil with regard to chemical composition and physical characteristics. Among the ingredients made from jojoba, simmondsia chinensis (jojoba) seed oil is most frequently used in cosmetics and personal care products. Product categories in which simmondsia chinensis (jojoba) oil may be found include bath products, eye makeup, hair care products, makeup, nail care products, personal hygiene products, shaving products and skin care products.

Simmondsia chinensis (jojoba) seed oil is obtained by pressing the seed kernels of an evergreen shrub native to the sonoran and mojave deserts of Arizona, California and Mexico. This oil is different from other common plant oils in that it is composed almost completely (97%) of wax esters of monounsaturated, straight-chain acids and alcohols with high molecular weights. This makes jojoba oil and its derivative jojoba esters more similar to sebum and whale oil than to traditional vegetable oils. Jojoba is grown for the liquid jojoba oil in its seeds. Jojoba oil is easily refined to be odorless and colorless. Jojoba oil is also stable to oxidation. Jojoba (seed) oil and its derivatives are used primarily as hair conditioning agents and skin conditioning agents (occlusive).

The CIR Expert Panel evaluated the scientific data and based on the available information concluded that jojoba oil and the related ingredients were safe for use as cosmetic ingredients.

Key Words & Phrases   重点词汇

Ⅰ. Trehalose

trehalose['tri:hələus]　*n*. 海藻糖
molecular weight　分子量
melting point　熔点
dihydrate[dai'haidreit]　*n*. 二水合物；二水物
anhydrous[æn'haidrəs]　*adj*. 无水的
formamide[fɔ'mæmaid]　*n*. 甲酰胺
dimethyl sulfuoxide　二甲亚砜
alcohol['ælkəhɔ:l]　*n*. 酒精，乙醇
ether['i:θər]　*n*. 乙醚
unreactive[ˌʌnri'æktiv]　*adj*. 不起化学反应的；化学上惰性的
dissociate[di'sousieit]　*vt*. 游离；使分离；分裂　*vi*. 游离；分离；分裂
monosaccharidic[ˌmɒnəsæk'ærɪdɪk]　*n*. 单糖，单醣类（最简单的糖类）
constituent[kən'stitjuənt]　*n*. 成分；选民；委托人　*adj*. 构成的；选举的
extreme[ik'stri:m]　*adj*. 极端的；极度的；偏激的；尽头的　*n*. 极端；末端；最大程度；极端的事物
hydrolytic[ˌhaidrə'litik]　*adj*. 水解的，水解作用的

trehalase[trɪˈhɑlez]　*n*. 海藻糖酶
natural occurrence　自然发病；天然产状
insect[ˈɪnsekt]　*n*. 昆虫；卑鄙的人
ergot of rye[ˈɚgətʌvraɪ]　*n*. 麦角黑麦
starch[stɑːrtʃ]　*n*. 淀粉；刻板，生硬　*vt*. 给……上浆
enzyme[ˈenzaɪm]　*n*. 酶
maltooligosyl-trehalose synthase　*n*. 麦芽寡糖-海藻糖合酶
maltooligosyl-trehalose trehalohydrolase　*n*. 麦芽寡糖-海藻糖水解酶

Ⅱ. Ceramides

ceramide[ˈserəmaid]　*n*. 神经酰胺
miscellaneous[ˌmɪsəˈleɪniəs]　*adj*. 混杂的，各种各样的；多方面的，多才多艺的
naturally-occurring　自然发生的
epidermal[ˈɛpəˈdɚml]　*adj*. 表皮的；外皮的
epidermal permeability barrier　表皮渗透屏障
lipid[ˈlɪpɪd]　*n*. 脂质；油脂
sphingomyelin[ˌsfɪŋəʊˈmaɪəlin]　*n*. (神经)鞘磷脂
bilayer[ˈbaɪˌleɚ]　*n*. 双分子层(膜)
membranes[mɛmbrens]　*n*. 细胞膜(membrane 的复数)；薄膜；膜皮
stratum corneum[ˈkɔːniəm]　*n*. 角质层；角层；角化层
lipophilic[lɪpəˈfɪlɪk]　*adj*. 亲脂性的，亲脂的
hydroxypalmitoyl sphinganine　角羟棕榈酰二氢鞘氨醇，神经类鞘脂
2-oleamido-1,3-octadecanediol　2-油酰胺-1,3-十八烷二醇
bovine[ˈboʊvaɪn]　*adj*. 牛的；似牛的；迟钝的　*n*. 牛科动物
tissue[ˈtɪʃuː]　*n*. 纸巾；薄纱；一套　*vt*. 饰以薄纱；用化妆纸揩去
central nervous system tissue　中枢神经系统组织
spinal cord　*n*. 脊髓
transmit[trænzˈmɪt]　*vt*. 传输；传播；发射；传达；遗传　*vi*. 传输；发射信号
bovine spongiform encephalitis (bse)　*n*. 牛海绵状脑炎
designate[ˈdezɪgneɪt]　*vt*. 指定；指派；标出　*adj*. 指定的；选定的
phytosphingosines[faɪtoʊˈsfɪŋəsin]　*n*. 植物鞘氨醇

Ⅲ. Simmondsia Chinensis (Jojoba) Seed Oil

simmondsia chinensis (jojoba) seed oil　*n*. 荷荷巴油
shrub[ʃrʌb]　*n*. 灌木；灌木丛
desert shrub　*n*. 荒漠灌丛
hydrogenated[haɪˈdrɑːdʒəneɪtɪd]　*adj*. (油类)氢化的，加氢的
hydrolyzed[ˈhaɪdrəˌlaɪzd]　*adj*. 水解的
isomerized[aɪˈsɒməˌraɪzd]　*adj*. 异构化的
kernel[ˈkɜːrnl]　*n*. 核心，要点；内核；仁；麦粒，谷粒；精髓

evergreen shrub *n.* 常绿灌木

monounsaturated[ˌmɒnəʌnˈsætʃəreɪtɪd] *adj.* 单不饱和的

sebum[ˈsiːbəm] *n.* 皮脂；牛羊脂

whale oil *n.* 鲸油；鱼油；鲸脂油

oxidation[ˌɑːksɪˈdeɪʃn] *n.* 氧化

occlusive[əˈkluːsɪv] *adj.* 咬合的；闭塞的 *n.* 闭塞音

Practical Activities 实践活动

Part A Reading comprehension.

1. According to the passage, sphingomyelin (   ).

A. contains ceramides

B. a major component of the lipid bilayer that forms cell membranes

C. a component in the stratum corneum

D. all of the above

2. According to the passage, the oral LD_{50} was greater than (   ) mg/kg in rat studies of ceramide 2.

A. 200   B. 2000   C. 5000   D. none of the above

3. According to the passage, which statement in the following is correct? (   )

A. Ceramides derived from natural sources are all permitted for use in cosmetic products.

B. Ceramides derived from bovine central nervous system tissues are not permitted for use in cosmetic products.

C. Ceramides derived from bovine central nervous system tissues are permitted for use in cosmetic products.

D. Ceramides derived from plant sources are not permitted for use in cosmetic products.

4. According to the passage, simmondsia chinensis (Jojoba) seed oil is composed of (   ).

A. 97% of wax esters of monounsaturated

B. straight-chain acids

C. alcohols with high molecular weights

D. all of the above

5. According to the passage, which statement in the following is correct? (   )

A. Jojoba Oil is similar to sebum and whale oil.

B. Jojoba Oil is similar to traditional vegetable oils.

C. Jojoba Oil is similar to wax esters of saturated.

D. Jojoba Oil is similar to straight-chain acids.

Part B Translation Exercise.

1. Very soluble in water, formamide, and dimethyl sulfuoxide; soluble hot alcohol; slightly soluble to insolubel in ether.

2. Ceramide ingredients function primarily as hair conditioning agents and skin conditioning agents-miscellaneous in cosmetics.

3. Synthetic Jojoba Oil is mixture of wax esters of fatty acids and alcohols that is indistinguishable

from natural jojoba oil with regard to chemical composition and physical characteristics.
4. Simmondsia Chinensis (Jojoba) Seed Oil is obtained by pressing the seed kernels.
5. Jojoba oil is also stable to oxidation.
6. Jojoba (seed) Oil and its derivatives are used primarily as hair conditioning agents and skin conditioning agents (occlusive).

知识拓展

Botanical Ingredients

Unit Four
Administration & Regulation
化妆品监管与法规

Learning Objectives 学习目标

In this unit you will be able to:
1. Have an overview of FDA's authority over cosmetics including FDA's Recall Policy, Voluntary Cosmetic Registration Program, and Cosmetics Labeling Regulations.
2. Have an overview of EU Cosmetics Regulation.
3. Learn about COLIPA.
4. Have an overview of China's Import and Trade Regulations Over Cosmetics including Hygiene License, Record-keeping Certificate & CIQ Labels of Imported Cosmetics.

Lesson 14 U.S. Cosmetics Laws & Regulations
美国化妆品法规

 Lead in 课前导入

Thinking and talking:
1. What is FDA's authority over cosmetics?
2. What is the definition of cosmetic by intended use?
3. What labeling information is required?

Related words:

import	进口
export	出口
Food and Drug Administration (FDA)	美国食品和药品管理局
Federal Food, Drug, and Cosmetic Act	联邦食品药品化妆品法案
adulterated or misbranded	掺假伪劣或错误标注
recall	召回
Voluntary Cosmetic Registration Program	化妆品自愿注册计划
labeling	标签

Task 1 FDA's Authority over Cosmetics

任务一 美国 FDA 对化妆品的监管

Ⅰ. FDA Mission

The Food and Drug Administration is responsible for protecting the public health by ensuring the safety, efficacy, and security of human and veterinary drugs, biological products, and medical devices; and by ensuring the safety of our nation's food supply, cosmetics, and products that emit radiation.

The Food and Drugs Act of 1906 was the first of more than 200 laws that constitute one of the world's most comprehensive and effective networks of public health and consumer protections. Today, the FDA regulates $1 trillion worth of products a year. It ensures the safety of all food except for meat, poultry and some egg products; ensures the safety and effectiveness of all drugs, biological products (including blood, vaccines and tissues for transplantation), medical devices, and animal drugs and feed; and makes sure that cosmetics and medical and consumer products that emit radiation do no harm.

Ⅱ. FDA Authority over Cosmetics

The two most important laws pertaining to cosmetics marketed in the United States are the Federal Food, Drug, and Cosmetic Act (FD & C Act) and the Fair Packaging and Labeling Act (FPLA). FDA regulates cosmetics under the authority of these laws.

The FD & C Act defines cosmetics by their intended use, as "articles intended to be rubbed, poured, sprinkled, or sprayed on, introduced into, or otherwise applied to the human body...for cleansing, beautifying, promoting attractiveness, or altering the appearance" [FD & C Act, sec. 201(i)]. Among the products included in this definition are skin moisturizers, perfumes, lipsticks, fingernail polishes, eye and facial makeup, cleansing shampoos, permanent waves, hair colors, and deodorants, as well as any substance intended for use as a component of a cosmetic product. It does not include soap.

The FD & C Act prohibits the marketing of adulterated or misbranded cosmetics in interstate commerce. "Adulteration" refers to violations involving product composition—whether they result from ingredients, contaminants, processing, packaging, or shipping and handling. "Misbranding" refers to violations involving improperly labeled or deceptively packaged products.

Under the FD & C Act, a cosmetic is adulterated if
- "it bears or contains any poisonous or deleterious substance which may render it injurious to users under the conditions of use prescribed in the labeling thereof, or under conditions of use as are customary and usual" (with an exception made for coal-tar hair dyes);

Lesson 14 U. S. Cosmetics Laws & Regulations

- "it consists in whole or in part of any filthy, putrid, or decomposed substance";
- "it has been prepared, packed, or held under insanitary conditions whereby it may have become contaminated with filth, or whereby it may have been rendered injurious to health";
- "its container is composed, in whole or in part, of any poisonous or deleterious substance which may render the contents injurious to health"; or
- except for coal-tar hair dyes, "it is, or it bears or contains, a color additive which is unsafe within the meaning of section 721(a)" of the FD & C Act (FD & C Act, sec. 601).

Under the FD & C Act, a cosmetic is misbranded if

- "its labeling is false or misleading in any particular";
- its label does not include all required information. (An exemption may apply to cosmetics that are to be processed, labeled, or repacked at an establishment other than where they were originally processed or packed; (see Title 21, Code of Federal Regulations, section 701.9.)
- the required information is not adequately prominent and conspicuous;
- "its container is so made, formed, or filled as to be misleading";
- it is a color additive, other than a hair dye, that does not conform to applicable regulations issued under section 721 of the FD & C Act; and
- "its packaging or labeling is in violation of an applicable regulation issued pursuant to section 3 or 4 of the Poison Prevention Packaging Act of 1970." (FD & C Act, sec. 602)

Under the FD & C Act, a product also may be misbranded due to failure to provide material facts. This means, for example, any directions for safe use and warning statements needed to ensure a product's safe use.

In addition, under the authority of the FPLA, FDA requires a list of ingredients for cosmetics marketed on a retail basis to consumers (Title 21, CFR, section 701.3). Cosmetics that fail to comply with the FPLA are considered misbranded under the FD & C Act (FPLA, section 1456). This requirement does not apply to cosmetics distributed solely for professional use, institutional use (such as in schools or the workplace), or as free samples or hotel amenities.

FDA's legal authority over cosmetics is different from our authority over other products we regulate, such as drugs, biologics, and medical devices. Under the law, cosmetic products and ingredients do not need FDA premarket approval, with the exception of color additives. However, FDA can pursue enforcement action against products on the market that are not in compliance with the law, or against firms or individuals who violate the law.

FDA does not pre-approve cosmetic products or ingredients, with the important exception of color additives. However, cosmetic firms are responsible for marketing safe, properly labeled products; using no prohibited ingredients; and adhering to limits on restricted ingredients.

Before marketing a product containing a color additive in the United States, it is essential to determine whether the additive is approved for its intended use. A number of color additives must be certified for purity in FDA labs if they are to be used legally in a product marketed in the United States.

Although U. S. regulations do not specify any particular testing regimens for cosmetic products or ingredients, it is the cosmetic company's responsibility to substantiate product and ingredient safety prior to marketing. In general, except for color additives and those ingredients that are prohibited or restricted by regulation, a manufacturer may use any ingredient in the formulation of a cosmetic, provided that

- the ingredient and the finished cosmetic are safe under labeled or customary conditions of use
- the product is properly labeled, and
- the use of the ingredient does not otherwise cause the cosmetic to be adulterated or misbranded under the laws that FDA enforces.

FDA-regulated does not mean FDA-approved. FDA does not have the legal authority to approve cosmetics before they go on the market, although we do approve color additives used in them (except coal tar hair dyes). However, under the law, cosmetics must not be "adulterated" or "misbranded." For example, they must be safe for consumers when used according to directions on the label, or in the customary or expected way, and they must be properly labeled. Companies and individuals who market cosmetics have a legal responsibility for the safety and labeling of their products. FDA can take action against a cosmetic on the market if we have reliable information showing that it is adulterated or misbranded. For example, FDA can pursue action through the Department of Justice in the federal court system to remove adulterated and misbranded cosmetics from the market. To prevent afurther shipment of an adulterated or misbranded product, FDA may request a federal district court to issue a restraining order against the manufacturer or distributor of the violative cosmetic. Cosmetics that are not in compliance with the law may be subject to seizure.

Key Words & Phrases　重点词汇

Food and Drug Administration (FDA)　美国食品和药品管理局
veterinary['vetrənəri]　*adj.* 兽医的
radiation[,reɪdɪ'eɪʃn]　*n.* 辐射；发光；放射物
comprehensive[,kɒmprɪ'hensɪv]　*adj.* 综合的；广泛的；有理解力的　*n.* 综合学校；专业综合测验
trillion['trɪljən]　*num.* 万亿
poultry['pəʊltri]　*n.* 家禽，家禽肉
vaccine['væksiːn]　*n.* 疫苗；牛痘苗
tissues for transplantation　组织移植
emit radiation　释放辐射
pertaining[pə'teɪnɪŋ]　*adj.* 附属的；与……有关的　*n.* 关于(pertain 的 ing 形式)
Federal Food, Drug, and Cosmetic Act　联邦食品药品化妆品法案
Fair Packaging and Labeling Act　正确包装和标识法案

Lesson 14　U. S. Cosmetics Laws & Regulations

intended use　预期用途
rubbed[rʌbd]　*v*. 擦（rub 的过去式和过去分词）；摩擦；搓
sprinkle['sprɪŋkl]　*v*. 撒，洒
permanent wave　烫发
deodorant[di:'əʊdərənt]　*n*. 除臭剂
adulterated or misbranded　掺假伪劣或错误标注
interstate commerce　州际贸易
adulteration[ə,dʌltə'reɪʃn]　*n*. 掺杂；搀假货
violation[,vaɪə'leɪʃn]　*n*. 违反；妨碍，侵害
deleterious[,delə'tɪəriəs]　*adj*. 有毒的，有害的
coal-tar['kəʊl tɑːr]　*n*. 煤焦油
filthy['fɪlθi]　*adj*. 肮脏的；污秽的
putrid['pjuːtrɪd]　*adj*. 腐败的；腐烂的
insanitary[ɪn'sænətri]　*adj*. 不卫生的；有害健康的
exemption[ɪg'zempʃn]　*n*. 免除，豁免；免税
prominent['prɒmɪnənt]　*adj*. 突出的，显著的；杰出的；卓越的
conspicuous[kən'spɪkjuəs]　*adj*. 显著的；显而易见的
Poison Prevention Packaging Act　防止有毒物包装法案
solely['səʊlli]　*adv*. 单独地；仅仅
institutional[,ɪnstɪ'tjuːʃənl]　*adj*. 制度的；公共机构的
hotel amenities　酒店设施
legally['liːgəli]　*adv*. 合法地；法律上
Department of Justice　司法部
restraining order　禁令
seizure['siːʒə(r)]　*n*. 没收；夺取；捕获

Practical Activities　实践活动

Reading comprehension.

1. According to the passage, which product is NOT responsible for protecting the public health by ensuring the safety?（　　）
A. human and animal drugs　　　　　B. biological products
C. cosmetic products　　　　　　　　D. meat, poultry and some egg products

2. Which product is NOT included in the FD & C Act by its intended use?（　　）
A. soap　　　　　　　　　　　　　　B. permanent waves
C. deodorants　　　　　　　　　　　D. perfumes

3. Under the FD & C Act, it is NOT adulterated if a cosmetic（　　）.
A. bears or contains any poisonous or deleterious substance
B. consists in whole or in part of any filthy, putrid, or decomposed substance
C. its container is composed of any poisonous or deleterious substance
D. a color additive which is safe within the FD & C Act

4. Under the FD & C Act, it is NOT misbranded if a cosmetic (   ).

A. its labeling is false or misleading in any particular

B. its label does not include all required information

C. its container is so made, formed, or filled as to be misleading

D. its packaging or labeling conform to the Poison Prevention Packaging Act

5. Under the authority of the FPLA, which of the following product requires a list of ingredients for cosmetics? (   )

A. professional use   B. institutional use

C. free samples or hotel amenities   D. aretail basis to consumers

6. A manufacturer may use any ingredient in the formulation of a cosmetic, provided that except for? (   )

A. the ingredient and the finished cosmetic that are safe under labeled or customary conditions of use

B. the product that is properly labeled

C. color additives and those ingredients that are prohibited or restricted by regulation

D. the use of the ingredient that does not otherwise cause the cosmetic to be adulterated or misbranded under the laws that FDA enforces

知识拓展

食品和药品管理局（FDA）禁止和限制在化妆品中使用的成分

Task 2　FDA Recall Policy for Cosmetics
任务二　美国 FDA 对化妆品的召回政策

A recall is a firm's removal or correction of a marketed product that FDA considers to be in violation of the laws we administer and against which we would initiate legal action.

Ⅰ. What Is FDA's Role in a Recall?

FDA has no authority under the FD & C Act to order a recall of a cosmetic, although it can request that a firm recall a product. However, we do have an active role in recalls. For example: we monitor the progress of a recall. In addition to reviewing firm status reports, we may conduct our own audit checks at wholesale or retail customers to verify the recall's effectiveness. We evaluate the health hazard presented by the product under recall and assign a classification to indicate the degree of hazard posed by a product under recall. Class I is a situation in which there is a reasonable probability that the use of, or exposure to, a violative product will cause serious adverse health consequences or death. Class II is a situation in which use of, or exposure to, a violative product may cause temporary or medically reversible adverse health consequences or where the probability of serious adverse health consequences is remote. Class III is a situation in which use of, or exposure to, a violative product is not likely to cause adverse health consequences. If we believe that public notification is necessary, we assure that either FDA or the firm issues the public notification. If the firm is unwilling to issue a press release, or unduly delays issuing a press release, we will issue one. FDA also issues general information about recalls through a weekly publication, the FDA Enforcement Report, which provides information on all recalls that have been assigned a classification. If we request a recall, we develop a recommended strategy for each recall that sets forth how the agency expects it to be carried out and the necessity for any press release. If the firm develops a recall strategy, we review and comment on that strategy. We make sure that the product is destroyed or suitably reconditioned.

II. What Is a Cosmetic Firm's Responsibility in a Recall?

Under the guidelines in 21 CFR Part 7, you are expected to do the following: You should notify your customers. The content, format, and extent of notification should be commensurate with the hazard presented by the product and the recall strategy developed for the product, as detailed in 21 CFR 7.49. When you initiate a recall, you should notify the appropriate FDA district office. You should submit periodic recall status reports to the appropriate FDA district office so that we may assess the progress of the recall. If FDA or your firm determines that a public warning is necessary, you should submit such a statement and plan for its distribution to FDA for review and comment. You should conduct effectiveness checks, as described in 21 CFR 7.42(b)(3). You are responsible for the disposition of the recalled product, whether the product is destroyed or brought into compliance.

Key Words & Phrases 重点词汇

audit['ɔːdɪt] n. 审计,稽核；查账；审查,检查
reversible[rɪ'vɜːsəbl] adj. 可逆的；可撤消的；可反转的
press release 新闻稿；通讯稿
unduly[ʌn'djuːli] adv. 过度地；不适当地；不正当地
set forth 陈述,提出；出发；陈列；宣布
be commensurate with 与……相应；与……相当
initiate[ɪ'nɪʃieɪt] vt. 开始,创始；发起；使初步了解

submit[səb'mɪt] *vt.* 使服从；主张；呈递；提交
periodic[ˌpɪəri'ɒdɪk] *adj.* 周期的；定期的
disposition[ˌdɪspə'zɪʃn] *n.* 处置
compliance[kəm'plaɪəns] *n.* 顺从，服从；符合

Practical Activities　实践活动

Reading comprehension.

According to the health hazard presented by the product under recall, which degree of hazard is NOT classified into? （　　）
A. Class Ⅰ　　　B. Class Ⅱ　　　C. Class Ⅲ　　　D. Class Ⅳ

知识拓展

Task 3　FDA's Voluntary Cosmetic Registration Program（VCRP）

任务三　美国 FDA 的化妆品自愿注册计划

Cosmetic companies may register in the United States through FDA's Voluntary Cosmetic Registration Program（VCRP）. The VCRP assists FDA in carrying out its responsibility to regulate cosmetics. FDA uses the information to evaluate cosmetic products on the market.

Ⅰ. What is FDA's role in VCRP?

VCRP is a reporting system for use by manufacturers, packers, and distributors of cosmetic products that are in commercial distribution in the United States.

Under the law, manufacturers are not required to register their cosmetic establishments or file their product formulations with FDA, and no registration number is required to import cosmetics into the United States. However, FDA encourages cosmetic firms to participate in VCRP using the online registration system. Cosmetic manufacturers, distributors, and packers can file information on their products that are currently being marketed to consumers in the United States and register their manufacturing and/or packaging facility locations in the VCRP database.

Ⅱ. Benefits of VCRP Participation

The VCRP assists FDA in carrying out its responsibility to regulate cosmetics marketed in the United States. Because product filings and establishment registrations are not mandatory, voluntary submissions provide FDA with the best estimate of information available about cosmetic products and ingredients, their frequency of use, and businesses engaged in their manufacture and distribution.

Information from the VCRP database also has been used by the Cosmetic Ingredient Review (CIR), an independent, industry-funded panel of scientific experts, to assist the CIR Expert Panel in establishing their priorities for assessing ingredient safety as part of their ingredient safety review.

Key Words & Phrases 重点词汇

Voluntary Cosmetic Registration Program 化妆品自愿注册计划
establishment registration 企业注册
mandatory['mændətəri] *adj.* 强制的；托管的；命令的
submission[səb'mɪʃn] *n.* 投降；提交(物)；服从
Cosmetic Ingredient Review (CIR) 化妆品原料评价委员会

Practical Activities 实践活动

Reading comprehension.

which of the following might NOT register in FDA's VCRP？（   ）
A. cosmetic manufacturers　　　　B. distributors
C. packers　　　　　　　　　　　D. cosmetic consumers

知识拓展

Task 4 Cosmetics Labeling Regulations

任务四 化妆品标签管理规定

Overview of Labeling Requirements

The following information is a brief introduction to labeling requirements. For a more thorough explanation of cosmetic labeling regulations, refer to FDA's Cosmetic Labeling Guide and the cosmetic labeling regulations themselves (21 CFR parts 701 and 740). Firms also may wish to discuss their labeling needs with a consultant.

Proper labeling is an important aspect of putting a cosmetic product on the market. FDA regulates cosmetic labeling under the authority of both the Federal Food, Drug, and Cosmetic Act (FD & C Act) and the Fair Packaging and Labeling Act (FPLA). These laws and their related regulations are intended to protect consumers from health hazards and deceptive practices and to help consumers make informed decisions regarding product purchase.

It is illegal to introduce a misbranded cosmetic into interstate commerce, and such products are subject to regulatory action. Some of the ways a cosmetic can become misbranded are:

- its labeling is false or misleading,
- its label fails to provide required information,
- its required label information is not properly displayed, and
- its labeling violates requirements of the Poison Prevention Packaging Act of 1970 [FD & C Act, sec. 602; 21 U.S.C. 362].

Does FDA Pre-approve Cosmetic Product Labeling?

No. FDA does not have the resources or authority under the law for pre-market approval of cosmetic product labeling. It is the manufacturer's and/or distributor's responsibility to ensure that products are labeled properly. Failure to comply with labeling requirements may result in a misbranded product.

Some Labeling Terms to Know

Before proceeding with a discussion of labeling requirements, it is helpful to know what some labeling terms mean:

- Labeling. This term refers to all labels and other written, printed, or graphic matter on or accompanying a product [FD & C Act, sec. 201(m); 21 U.S.C. 321(m)].
- Principal Display Panel (PDP). This is the part of the label most likely displayed or examined under customary conditions of display for sale [21 CFR 701.10].
- Information Panel. Generally, this term refers to a panel other than the PDP that can accommodate label information where the consumer is likely to see it. Since the information must be prominent and conspicuous [21 CFR 701.2(a) (2)], the bottom of the package is generally not acceptable for placement of required information, such as the cosmetic ingredient declaration.

Is It Permitted to Label Cosmetics "FDA Approved"?

No. As part of the prohibition against false or misleading information, no cosmetic may be labeled or advertised with statements suggesting that FDA has approved the product. This applies even if the establishment is registered or the product is on file with FDA's Voluntary Cosmetic Registration Program (VCRP) (see 21 CFR 710.8 and 720.9, which prohibit the use of participation in the VCRP to suggest official approval). False or misleading statements on labeling make a cosmetic misbranded [FD & C Act, sec. 602; 21 U.S.C. 362].

What About Therapeutic Claims?

Be aware that promoting a product with claims that it treats or prevents disease or otherwise affects the structure or any function of the body may cause the product to be considered a drug. FDA has an Import Alert in effect for cosmetics labeled with drug claims. For more information on drug claims, refer to *Is It a Drug, a Cosmetic, or Both?*

How Should Products Be Labeled If They Are Both Drugs and Cosmetics?

If a product is an over-the-counter (OTC) drug as well as a cosmetic, its labeling must comply with the regulations for both OTC drug and cosmetic ingredient labeling [21 CFR 701.3 (d)]. The drug ingredients must appear according to the OTC drug labeling requirements [21 CFR 201.66(c) (2) and (d)] and the cosmetic ingredients must appear separately, in order of decreasing predominance [21 CFR 201.66 (c) (8) and (d)]. Contact the Center for Drug Evaluation and Research (CDER) for further information on drug labeling.

What Languages Are Acceptable?

All labeling information that is required by law or regulation must be in English. The only exception to this rule is for products distributed solely in a U.S. territory where a different language is predominant, such as Puerto Rico. If the label or labeling contains any representation in a foreign language, all label information required under the FD & C Act must also appear in that language [21 CFR 701.2(b)]. For information on dual declaration of ingredients, see "Ingredient Names".

What Labeling Information Is Required?

The following information must appear on the principal display panel:

- An identity statement, indicating the nature and use of the product, by means of either the common or usual name, a descriptive name, a fanciful name understood by the public, or an illustration [21 CFR 701.11].
- An accurate statement of the net quantity of contents, in terms of weight, measure, numerical count or a combination of numerical count and weight or measure [21 CFR 701.13].

The following information must appear on an information panel:

- Name and place of business. This may be the manufacturer, packer, or distributor. This includes the street address, city, state, and ZIP code. You may omit the street address if it is listed in a current phone directory or city directory [21 CFR 701.12(a)].
- Distributor statement. If the name and address are not those of the manufacturer, the label must say "Manufactured for..." or "Distributed by..." or similar wording expressing the facts [21 CFR 701.12(c)].
- Material facts. Failure to reveal material facts is one form of misleading labeling and therefore makes a product misbranded [21 CFR 1.21]. An example is directions for safe use, if a product could be unsafe if used incorrectly.
- Warning and caution statements. These must be prominent and conspicuous. The FD & C Act and related regulations specify warning and caution statements related to specific products [21 CFR part 700]. In addition, cosmetics that may be hazardous to consumers must bear appropriate label warnings [21 CFR 740.1]. An example of such hazardous products is flammable cosmetics.

- Ingredients. If the product is sold on a retail basis to consumers, even it is labeled "For professional use only" or words to that effect, the ingredients must appear on an information panel, in descending order of predominance. [21 CFR 701.3]. Remember, if the product is also a drug, its labeling must comply with the regulations for both OTC drug and cosmetic ingredient labeling, as stated above. To learn more, see "Ingredient Names" "Color Additives and Cosmetics" "Fragrances in Cosmetics" and "Trade Secret' Ingredients"

Key Words & Phrases 重点词汇

thorough[ˈθʌrə] *adj.* 彻底的；十分的；周密的
deceptive[dɪˈseptɪv] *adj.* 欺诈的；迷惑的；虚伪的
illegal[ɪˈliːgl] *adj.* 非法的；违法的；违反规则的
accompanying[əˈkʌmpəniɪŋ] *adj.* 伴随的
Principal Display Panel (PDP) 主要展示版面
accommodate[əˈkɒmədeɪt] *vt.* 容纳；使适应；供应；调解
therapeutic[ˌθerəˈpjuːtɪk] *adj.* 治疗的；治疗学的；有益于健康的
over-the-counter (OTC)[ˌəʊvə ðə ˈkaʊntə(r)] 非处方的
predominance[prɪˈdɒmɪnəns] *n.* 优势；卓越
Center for Drug Evaluation and Research(CDER) 药品评价与研究中心
Puerto Rico[ˈpwɜːtəʊˈriːkəʊ] 波多黎各
net quantity 净含量
numerical count 数码；数字码
ZIP code 邮政编码
omit[əˈmɪt] *vt.* 省略；遗漏；删除；疏忽
Material facts 重要事实
hazardous[ˈhæzədəs] *adj.* 有危险的；冒险的；碰运气的
flammable[ˈflæməbl] *adj.* 易燃的；可燃的；可燃性的
fragrances[ˈfreɪɡrəns] *n.* 香精
trade secret 商业秘密

Practical Activities 实践活动

Part A Reading comprehension.

1. Which of the following information must NOT appear on an information panel?（　　）
   A. distributor statement   B. the street address of business
   C. warning and caution statements   D. ingredients
2. Which of the terms a product's labeling do not include?（　　）
   A. written matter   B. printed matter
   C. graphic matter   D. FDA Approved

Lesson 14 U. S. Cosmetics Laws & Regulations

Part B Matching the terms of the following regulations.

1. 联邦食品药品化妆品法案	1. Fair Packaging and Labeling Act
2. 正确包装和标识法案	2. Federal Food, Drug, and Cosmetic Act
3. 防止有毒物包装法案	3. interstate commerce
4. 司法部	4. Food and Drug Administration
5. 美国食品和药品管理局	5. Department of Justice
6. 州际贸易	6. Poison Prevention Packaging Act
7. 化妆品自愿注册计划	7. Principal Display Panel
8. 化妆品原料评价委员会	8. over-the-counter
9. 主要展示版面	9. Cosmetic Ingredient Review
10. 非处方的	10. Voluntary Cosmetic Registration Program
11. 药品评价与研究中心	11. Center for Drug Evaluation and Research

知识拓展

Cosmetics Labeling

Lesson 15 EU Cosmetics Regulation
欧盟化妆品法规

Lead in 课前导入

Thinking and talking:
1. How many kinds of the EU's cosmetics legislation do you know?
2. What's the EU's cosmetic products notification portal?
3. What's Colipa and its membership consist of?

Related words:

internal	国内的
foreign	国外
European Commission (EC)	欧盟委员会
European Union (EU)	欧洲联盟
dominant	占优势的
Colipa	欧洲化妆品行业协会
nanomaterial	纳米材料
cosmetic products notification portal (CPNP)	化妆品备案门户

Task 1 Regulation (EC) No 1223/2009
任务一 欧盟化妆品管理法规 No 1223/2009

Ⅰ. The EU's Role in Cosmetics

Cosmetics range from everyday hygiene products such as soap, shampoo, deodorant, and toothpaste to luxury beauty items including perfumes and makeup. These products are regulated at European level to ensure consumers' safety and to secure an internal market for cosmetics.

Europe is a world leader in the cosmetics industry and dominant cosmetics exporter. The sector is highly innovative and provides significant employment in Europe. The EU's involvement mainly concerns the regulatory framework for market access, international trade relations, and regulatory convergence. These all aim to ensure the highest level of consumer safety while promoting the innovation and the competitiveness of this sector.

The European Commission is also in contact with cosmetics stakeholders at EU and international level. This cooperation enables the exchange of information and ensures the smoother implementation of EU requirements in the sector.

II. Product Safety and Legislation

Regardless of the manufacturing processes or the channels of distribution, cosmetic products placed on the EU market must be safe. The manufacturer is responsible for the safety of their products, and must ensure that they undergo an expert scientific safety assessment before they are sold. A special database with information on cosmetic substances and ingredients, called CosIng, enables easy access to data on these substances, including legal requirements and restrictions.

Cosmetics legislation at EU level also
- requires that all products to be marketed in the EU must be registered in the cosmetic products notification portal (CPNP) before being placed on the market;
- requires that some cosmetic products are given special attention from regulators due to their scientific complexity or higher potential risk to consumer health;
- ensures that there is a ban on animal testing for cosmetic purposes;
- makes EU countries responsible for market surveillance at national level.

III. Legislation

Regulation (EC) No 1223/2009 on cosmetic products is the main regulatory framework for finished cosmetic products when placed on the EU market. It strengthens the safety of cosmetic products and streamlines the framework for all operators in the sector. The regulation simplifies procedures to the extent that the internal market of cosmetic products is now a reality. The regulation replaces Directive 76/768/EC, which was adopted in 1976 and had been substantially revised on numerous occasions. It provides a robust, internationally recognised regime, which reinforces product safety while taking into consideration the latest technological developments, including the possible use of nanomaterials. The previous rules on the ban of animal testing were not modified.

IV. Other Applicable EU Legislation

Apart from the main regulatory framework for finished cosmetic products, additional requirements covered by other EU legislation might apply. Some of them are listed below.
- Chemicals website of the European Commission including information on restrictions on the marketing and use of certain dangerous substances and preparations.
- Climate website of the European Commission regarding ozone-depleting substances con-

tained in aerosol products.
- Pressure and gas website of the European Commission relating to aerosol dispensers.
- Legal metrology and pre-packaging website of the European Commission with information on nominal quantities and nominal capacities permitted for certain prepackaged products.
- Environment website of the European Commission on packaging and packaging waste.

The most significant changes introduced by the cosmetics regulation include.
- Strengthened safety requirements for cosmetic products

  Manufacturers need to follow specific requirements in the preparation of a product safety report prior to placing a product on the market.
- Introduction of the notion of "responsible person"

  Only cosmetic products for which a legal or natural person is designated within the EU as a "responsible person" can be placed on the market. The new cosmetics regulation allows the precise identification of the responsible person is and clearly outlines their obligations.
- Centralised notification of all cosmetic products placed on the EU market.

  Manufacturers will need to notify their products only once—via the EU cosmetic products notification portal (CPNP).
- Introduction of reporting of serious undesirable effects (SUE)

  A responsible person will have an obligation to notify serious undesirable effects to national authorities. The authorities will also collect information coming from users, health professionals, and others. They will be obliged to share the information with other EU countries. More information on reporting of SUE.
- New rules for the use of nanomaterials in cosmetic products

  Colourants, preservatives and UV-filters, including those that are nanomaterials, must be explicitly authorised. Products containing other nanomaterials not otherwise restricted by the cosmetics regulation will be the object of a full safety assessment at EU level if the commission has concerns. Nanomaterials must be labelled in the list of ingredients with the word "nano" in brackets following the name of the substance, e.g. "titanium dioxide (nano)".

Ⅴ. Cosmetic Product Notification Portal

The cosmetic products notification portal (CPNP) is a free of charge online notification system created for the implementation of Regulation (EC) No 1223/2009 on cosmetic products. When a product has been notified in the CPNP, there is no need for any further notification at national level within the EU.

Regulation (EC) No 1223/2009 requires that the responsible persons and, under certain circumstances, the distributors of cosmetic products submit some information about the products they place or make available on the European market through the CPNP.

The CPNP is making this information available electronically to
- competent authorities (for the purposes of market surveillance, market analysis, evaluation and consumer information);

- poison centres or similar bodies established by EU countries (for the purposes of medical treatment).

The CPNP is accessible to
- competent authorities;
- European poison centres;
- cosmetic products responsible persons;
- distributors of cosmetic products.

Key Words & Phrases 重点词汇

European Commission　欧盟委员会
European Union　欧洲联盟
hygiene['haɪdʒiːn]　n. 卫生
deodorant[diː'əʊdərənt]　n. 除臭剂
innovative['ɪnəveɪtɪv]　adj. 革新的，创新的；新颖的
stakeholders['steɪkhəʊldə(r)]　n. 利益相关者；赌金保管者
restrictions[rɪ'strɪkʃn]　n. 限制；约束；束缚
cosmetic products notification portal (CPNP)　化妆品备案门户
surveillance[sɜː'veɪləns]　n. 监督；监视
on numerous occasions　无数次
robust[rəʊ'bʌst]　adj. 强健的；健康的；粗野的；粗鲁的
take into consideration　考虑到……
nanomaterial　纳米材料
ozone-depleting　消耗臭氧层的
aerosol['eərəsɒl]　n. 气溶胶；气雾剂
legal metrology　法制计量
nominal quantities　标称数量
nominal capacities　额定容量；公称容积
obligations[ˌɒblɪ'geɪʃn]　n. 义务；职责；债务
centralise['sentrəlaɪz]　vt. 把……集中起来；形成中心
serious undesirable effects (SUE)　严重不良影响
colourants　着色剂
preservatives　防腐剂
explicitly[ɪk'splɪsɪtli]　adv. 明确地；明白地
authorized['ɔːθəraɪzd]　adj. 授权的　v. 授权
brackets['brækɪts]　n. 支架；括号，圆括号
titanium dioxide　二氧化钛
free of charge　免费
under certain circumstances　adj. 有时；在某种情况下（在某种状况下）
competent authorities　主管部门；主管当局
European poison centres　欧洲毒物控制中心

Practical Activities 实践活动

Reading comprehension.

1. According to the passage, which part is Not the EU's involvement mainly concerns?(　　)
   A. market access
   B. international trade relations
   C. regulatory convergence
   D. cosmetic products

2. According to the passage, cosmetics legislation at EU level also(　　).
   A. don't require that all products to be marketed in the EU must be registered in the cosmetic products notification portal (CPNP) before being placed on the market
   B. don't require that some cosmetic products are given special attention from regulators due to their scientific complexity or higher potential risk to consumer health
   C. don't ensure that there is a ban on animal testing for cosmetic purposes
   D. makes EU countries responsible for market surveillance at national level

3. Which of the following EU legislation might NOT apply?(　　)
   A. certain dangerous substances and preparations
   B. ozone-depleting substances contained in aerosol products
   C. directive 76/768/EC
   D. packaging and packaging waste

4. According to the passage, the most significant changes introduced by the cosmetics regulation DON'T include?(　　)
   A. strengthened safety requirements for cosmetic products
   B. introduction of the notion of 'responsible person'
   C. introduction of reporting of serious undesirable effects (SUE)
   D. the ban of animal testing

5. The CPNP is NOT accessible to(　　).
   A. competent authorities
   B. European poison centres
   C. consumers
   D. cosmetic products responsible persons

知识拓展

Task 2　Colipa (European Cosmetics Trade Association)

任务二　欧洲化妆品行业协会

Colipa is the umbrella body in Europe that governs all cosmetic and product care products manufactured and later are on the market.

In achieving its goal, Colipa cannot work alone. As such it has incorporated different legal and private investors to propel its mission to greater heights. For example, it must provide sample products to quality control authorities so as to certify quality standards. Additionally, Colipa engages legal entities to ensure products are safe for use. Other authorities governing the industry are ISO, Food and Drug Administration and the European Union (EU) Cosmetic regulation.

As the industry's European trade association Colipa's membership consists of: international companies, national associations, supporting association members, supporting corporate members, correspondent members.

Who We Are

Colipa is the European trade association for cosmetic and personal care industry. Colipa represents the cosmetic industry at a national level where the association represents cosmetic and personal care manufacturers across Europe and beyond. The association represents a range number of innovative companies either directly or at national membership.

The need to have a product that is safe for the targeted group is important to both the producers and the consumers at large. Moreover, abiding by the set legislation in Europe requires that an umbrella body should be present to ensure quality. Individuals also like certified products that they feel safe when they use them. Over the last 50 years, Colipa has taken it on their shoulder. It ensures consumers get the best product available on the market. During this period the organization has been the voice of cosmetic and personal care industry in Europe.

The association follows regulations set at the European level. As such the association in collaboration with other policymakers, it also makes sure that European regulation is as effective and appropriate as it could be.

Our Role

The association ensures that the voice of the industry is heard in the EU legislative and policy arguments, and globally keeps the members informed of significant trends and developments relevant to the industry.

Mission

To ensure quality cosmetic and personal care products for consumers through innovative science. The primary goal of the association is to deliver a quality product on the market. This is done by the manufacturing of high-performance cosmetics and at the same time promotes innovation through research, improves efficiency, and remains relevant.

Vision

For a flourishing European Cosmetics and personal care industry.

Colipa thrives in its mandate to ensure that the products in the market are safe which in turn makes it flourish through an improved sale and maximizing profit. Making sure that industry gets the reputation it deserves is also in the visionary mind of its executives. Moreover, the association makes available different products which are safe for use and right for the industry.

As a member of the European trade association, both companies and associations help build the regulatory and policy picture under the jurisdiction of the association which the industry operates. Moreover, the association provides guidelines associated with environmental issues that must be met for the safety of the products in the market.

Colipa has the mandate to provide information and basic knowledge about the cosmetic; labeling is important as it provides the ingredients used in production which in turn boost the customer's confidence to use the product. Information about ingredient list, nominal net, warnings, date of minimum durability. Reference for product identification and country of origin must be on the label.

Cosmetic products have to be regulated to ensure their safety by the user. Stipulated regulations by the European Union's Cosmetic Regulation body stipulate how products should be produced and their safety. Each company should conform to set rules and keep its customer's safety first. Beyond set regulations, Colipa is mandated to inform customers of information that help in wise decision making on the use of cosmetic and personal care products. This can be made through marketing and communication channels.

All cosmetic and personal care products on the market in Europe are safe for use. In fact, it is the primary role of manufacturers to ensure that products released on the market have met thresholds for acceptable and safe products to consumers. Prior to the product being released in the market, the European Union legislation stipulates that all new products must be subjected to expect scientific safety assessment before they are launched for sale. For example, reduced rejection of products has been curbed by animal testing where prior testing assesses the side-effects of the product when used by humans. Such procedures are integral for better market reception of these products.

European Cosmetics Association

For a long time the association, Colipa, has been the European trade association for the cosmetic and personal care umbrella in Europe that covers all dynamics in the cosmetic industry in Europe. As a regulatory measure, the association is mandated to ensure correct follow up of rules governing the cosmetic industry in Europe.

Moreover, the association established over five decades ago works with over 10 multinational companies, over 20 national associations and over 3,000 SMEs that ensure the industry provides European consumers with products that comply with EU regulation and safety standards.

The European cosmetic association additionally provides the necessary labeling guidelines which stipulate the required thresholds for specific personal care products. The association states specific requirements like, contents information, the origin of the product, the minimum date of durability which is important in the delivery of safe products to the consumers. The association continually relies on innovations that enhance quality and efficiency. The association invests on an annual budget of over 1.27 billion Euros to provide safe, sustainable and innovative products. Moreover, this effort does not pass the glance of the consumer. In particular, the study approves that the consumer must provide unnoticeable effort by the customer. The association further provides for 86% efficiency on products and 87% on quality. Moreover, financial considerations take account for 68% of cosmetic and personal considerations in Europe.

Key Words & Phrases 重点词汇

Colipa(European Cosmetics Trade Association) 欧洲化妆品行业协会
umbrella body 协调组织
legal entities 法人实体
remain relevant 与时俱进
flourishing['flʌrɪʃɪŋ] *adj.* 繁荣的；繁茂的；盛行的
mandate['mændeɪt] *n.* 授权；命令，指令；委托管理；受命进行的工作 *vt.* 授权；托管
in turn 反过来；转而
reputation[,repju'teɪʃn] *n.* 名声，名誉；声望
jurisdiction[,dʒʊərɪs'dɪkʃn] *n.* 司法权，审判权，管辖权；权限，权力
boost[buːst] *vt.* 促进；增加
nominal net 净含量
date of minimum durability 保质期
stipulated['stɪpjuleɪt] *adj.* 规定的 *v.* 规定；保证
wise[waɪz] *adj.* 明智的；聪明的；博学的
rejection[rɪ'dʒekʃn] *n.* 抛弃；拒绝；被抛弃的东西
curbed[kɜːb] *adj.* 约束的 *v.* 抑制
side-effects 副作用
integral['ɪntɪɡrəl] *adj.* 积分的；完整的，整体的
multinational[,mʌltɪ'næʃnəl] *adj.* 跨国公司的；多国的
SMEs abbr. 中小企业(small and medium-size enterprise)
annual budget *n.* 年度预算

Practical Activities 实践活动

Part A Reading comprehension.

1. Which of the following is NOT Colipa's membership? (　　)
 A. international companies　　　　B. national associations
 C. supporting corporate members　　D. animal testing agencies
2. Which content of the cosmetic products labelling may not be marked? (　　)
 A. ingredient list　　　　　　　　B. nominal net
 C. date of minimum durability　　　D. responsible persons
3. According to the passage, how many financial considerations is taking account for about cosmetic and personal considerations in Europe? (　　)
 A. 86%　　　　　　　　　　　　B. 68%
 C. 87%　　　　　　　　　　　　D. 78%

Part B Matching the terms of the following regulations.

1. 欧盟委员会	1. SMEs (small and medium-size enterprise)
2. 欧洲联盟	2. European Commission
3. 化妆品备案门户	3. European poison centres
4. 严重不良影响	4. Colipa (European Cosmetics Trade Association)
5. 欧洲毒物控制中心	5. European Union
6. 欧洲化妆品行业协会	6. serious undesirable effects (SUE)
7. 中小企业	7. cosmetic products notification portal

知识拓展

欧盟化妆品法规 1223/2009 附录

Lesson 16 China's Import and Trade Regulations
中国进出口化妆品管理

Lead in 课前导入

Thinking and talking:
1. How many levels of standards in China do you know?
2. How many kinds of cosmetic products are there by the China's import regulations?
3. What's the Guangdong Cosmetics Safety Regulations' impact do you think about the cosmetics market? Why?

Related words:

import	进口	cross-border	跨境
export	出口	e-commerce	电子商务
expiry date	有效期	specialuse cosmetics	特殊用途化妆品
most-favoured-nation (MFN)	最惠国	non-specialuse cosmetics	非特殊用途化妆品

Task 1 Hygiene Licence of Imported Cosmetics

任务一 进口化妆品卫生许可证

The detailed rules for the Implementation of the Regulations on Cosmetics Hygienic Supervision stipulate that, when a cosmetic product is first imported into China, its foreign manufacturer or agent must obtain and complete an application form for Hygiene Licence of Imported Cosmetics from the relevant hygiene administration department. They must submit their applications directly to the hygiene administration department under the State Council. Once the application dossiers are received, the hygiene administration department will set up a cosmetics safety panel to inspect the product in question. Products that pass the inspection will be issued with an Approval Document for Hygiene Licence of Imported Cosmetics which is valid for four years. Application for renewal can be submitted four to six months before the approval document's expiry date, and attachment of relevant information is not required. Please refer to the website of the National Medical Products Administration (NMPA).

When assessing the mainland cosmetics market, foreign players should pay attention to the relevant standards adopted by mainland authorities. Under the Standardisation Law of the People's Republic of China (Revised Draft 2017), standards are classified into five levels, national, trade, local, organisation and enterprise standards, in order of descending precedence. National standards are further classified into compulsory and recommended standards, represented by standard codes GB and GB/T. Likewise, trade standards are recommended standards. The cosmetics sector, classified under the category of light industry, is represented by the standard codes of QB and QB/T. Local standards are recommended standards.

Enterprise standards are applicable within the respective enterprises. Industry players should refer to www.standardcn.com and the Standardisation Administration of China (SAC) website to look up the standards relevant to them.

On 1 July 2018, the State Council cut most-favoured-nation (MFN) tariffs on 1,449 items of imported daily consumer goods, including garment, shoes and hats as well as cosmetics and household appliances, among others. The average tariff rate for cosmetics such as skincare and haircare products was slashed from 8.4% to 2.9%. See table 1 for the current tariffs on selected cosmetics categories.

Table 1 Import tariffs of selected cosmetic products in 2018

HS Code	Description	Tariff Rate /%
33030000	Perfumes and toilet waters	3
33041000	Lip make-up preparations	5
33042000	Eye make-up preparations	5
33043000	Manicure or pedicure preparations	5
33049100	Powders, whether or not compressed	5
33049900	Others (including preparations for the care of the skin, suntan preparations, etc.)	1
33051000	Shampoos	2
33052000	Preparations for permanent waving	3
33053000	Hair lacquers	3
33059000	Others	3

Source: Revised MFN Tariff Schedule for Daily Consumer Goods, Ministry of Finance of the People's Republic of China, 2018.

Lesson 16 China's Import and Trade Regulations

The Instructions for Consumer Goods-General Labelling of Cosmetics stipulates that all locally produced cosmetics or imported cosmetic products registered for inspection and distribution on the mainland must truthfully indicate on the product package the standard Chinese names of all added ingredients in the product.

According to the Naming Requirements for Cosmetics, which was implemented on 5 February 2010, the name of a cosmetic product should be concise, easy to understand and in line with the customs of the Chinese language. It must not contain anything that may mislead or deceive consumers. The Cosmetics Naming Guidelines, which were issued to complement the naming requirements, provide a list of expressions allowed or prohibited when naming cosmetic products. Eleven types of expressions are forbidden for use in the names of cosmetic products: these include arbitrary expressions, such as "special effect" "total effect" "powerful effect" "miraculous effect" "super effect" "extraordinary" "skin renewing", or "wrinkle removing"; expressions that falsely claim a product is "absolutely natural"; expressions that explicitly or implicitly indicate the medical effect of a product, such as "anti-bacterial" "bacteria-inhibiting" "bacteria-removing" "detoxifying" "anti-allergic" "scar-removing" "hair-growing" "hair-regenerating" "fat-reducing" "fat-dissolving" "body-slimming" "face-slimming", and "leg-slimming"; medical jargon; and names of celebrities in the medical field, such as "Bian Que" "Hua Tuo" "Zhang Zhongjing" and "Li Shizhen".

In the 2015 version of the Safety and Technical Standards for Cosmetics consists of eight chapters, namely: "Summary" "Requirements on Cosmetics Prohibited and Restricted Ingredients" "Requirements on Permitted Ingredients" "Physical and Chemical Testing Methods" "Microbiological Testing Methods" "Toxicological Testing Methods" "Human Safety Testing Methods", and "Efficacy Evaluation Methods in Humans". In the publication, the general safety and technical requirements for cosmetics have been refined. Revisions have also been made to the lists regarding prohibited/restricted and permitted ingredients as well as to the physical and chemical testing methods in the cosmetics inspection and evaluation methods.

As stipulated in the former China Food and Drug Administration (CFDA)'s Circular on Matters Related to Cosmetics Production Permit, cosmetics produced on and after 1 July 2017 must use new packaging logos labelled with information on the corresponding cosmetics production permit. The "QS" logo is no longer in use.

In accordance with the Administrative Measures on the Inspection, Quarantine and Supervision of Import and Export of Cosmetics, which was revised in 2018 for cosmetics subject to hygiene licensing or state archival filing, the approval documents for hygiene licensing or a certificate of archival filing of imported cosmetics approved by the relevant supervisory department should be submitted. For cosmetics not subject to hygiene licensing or state archival filing, documents to be submitted include the relevant safety assessment data on materials with a potential safety risk, and certification allowing production and sale of the cosmetics in the country/territory of production or a certificate of origin. In addition, samples of a

Chinese label, a foreign label and its Chinese version are also required. For finished cosmetic products without sales packaging, information such as the name, quantity/weight, specification and origin should also be provided.

Since 10 November 2018, the pre-approval process for the first-time import of non-special-use cosmetics has been wholly replaced by a unified system of record-filing management across the mainland. The NMPA will no longer accept administrative approval applications related to any such products. The term 'non-special-use cosmetics' refers to general beauty products that don't have a specific application (such as the promotion of hair growth, hair colouring, hair perming, hair removal, breast re-shaping, fitness enhancement, odour reduction, freckle removal or protection from the sun). The manufacturer of the prospective import must appoint a mainland representative to complete the requisite record-filing procedures via the NMPA website and obtain an electronic record-filing certificate before importing any such products. Please refer to the NMPA website for details of the applicable laws and regulations.

Cosmetics imported via cross-border e-commerce retailing channels, such as some lip and eye make-up preparations, may be treated as goods imported for personal use if they are on the List of Cross-Border E-Commerce Retail Imports (2018 version) and meet the conditions stipulated in the policies and regulations for cross-border e-commerce retail import. The requirements for first-time import licence, registration or filing of goods do not apply. The revised list took effect on 1 January 2019.

The Announcement of the NMPA on Matters Concerning the Implementation of the Commitment-based Examination and Approval System for the Renewal of Administrative Approval for Special-Use Cosmetics, which came into effect on 30 June 2019, is aimed at improving the efficiency of the examination and approval of special-use cosmetics. Applicants needing to renew an approval for the import of special-use cosmetics must carry out comprehensive self-checks of the products concerned, six months before the expiry date of the approval. They also need to submit the report on the self-checks and the renewal application on the NMPA's examination and approval platform for special-use cosmetics, 30 working days before the expiry date. The renewal application will be approved within 15 working days if the relevant requirements are fulfilled. If renewal of the approval is not granted, products may not be produced or imported after the approval expires.

The Guangdong Cosmetics Safety Regulations, China's first local legislation regarding the safety supervision of cosmetics, governs the entire cosmetics market in Guangdong province. These regulations, which came into force on 1 July 2019, strictly control the uses of raw materials and the whole process of the production of cosmetics. They also clearly spell out the rules on labels and logos. E-commerce platforms for cosmetics are required to conduct real-name registration and verify the business qualifications of cosmetics vendors intending to use their services. Please refer to the website of the Guangdong Provincial Medical Products Administration for details of the applicable laws and regulations.

Lesson 16　China's Import and Trade Regulations

Key Words & Phrases　重点词汇

The Detailed Rules for the Implementation of the Regulations on Cosmetics Hygienic Supervision　化妆品卫生监督条例实施细则
stipulate['stɪpjuleɪt]　vi. 规定；保证
Hygiene Licence of Imported Cosmetics　进口化妆品卫生许可证
dossiers['dɒsieɪ]　n. 档案，卷宗；病历表册
renewal[rɪ'njuːəl]　n. 更新，恢复；复兴；补充；革新；续借；重申
expiry date　到期日；有效期限
attachment[ə'tætʃmənt]　n. 附件
National Medical Products Administration（NMPA）　国家药品监督管理局
Standardisation Law of the People's Republic of China　中华人民共和国标准化法
Standardisation Administration of China（SAC）　中国标准化管理委员会
most-favoured-nation（MFN）　最惠国
tariffs['tærɪf]　n. 关税
garment['gɑːmənt]　n. 衣服，服装；外表，外观
appliances[ə'plaɪəns]　n. 家用电器
slash[slæʃ]　vt. 猛砍；鞭打
manicure['mænɪkjʊə(r)]　n. 修指甲，美甲，指甲护理
pedicure['pedɪkjʊə(r)]　n. 修趾甲术；足部治疗
The Instructions for Consumer Goods-General Labelling of Cosmetics　消费品使用说明 化妆品通用标签
Naming Requirements for Cosmetics　化妆品命名要求
concise[kən'saɪs]　adj. 简明的，简洁的
The Cosmetics Naming Guidelines　化妆品命名指南
arbitrary['ɑːbɪtrəri]　adj. 任意的；武断的；专制的
miraculous[mɪ'rækjələs]　adj. 不可思议的，奇迹的
extraordinary[ɪk'strɔːdnri]　adj. 非凡的；特别的；离奇的；特派的
celebrity[sə'lebrəti]　n. 名人
Safety and Technical Standards for Cosmetics　化妆品安全技术标准
revision[rɪ'vɪʒn]　n. 修正；复习；修订本
China Food and Drug Administration（CFDA）　中国食品药品监督管理局
Administrative Measures on the Inspection，Quarantine and Supervision of Import and Export of Cosmetics　化妆品进出口检验检疫监督管理办法
state archival filing　国家备案
hair perming　烫发
freckle['frekl]　n. 雀斑；斑点
requisite['rekwɪzɪt]　adj. 必备的，必不可少的；需要的
record-filing　备案
cross-border['krɔːsbɔːrdər]　adj.（公司）跨国的

e-commerce[iːˈkɒmɜrs] n. 电子商务
List of Cross-Border E-Commerce Retail Imports 跨境电子商务零售进口商品清单
The Guangdong Cosmetics Safety Regulations 广东省化妆品安全条例
vendors[ˈvendə(r)] n. 供应商
Guangdong Provincial Medical Products Administration 广东省药品监督管理局

Practical Activities 实践活动

Part A Reading comprehension.

1. According to the passage, Approval Document for Hygiene Licence of Imported Cosmetics is valid for (　　) years?
   A. three　　　B. four　　　C. five　　　D. six

2. Under the Standardisation Law of the People's Republic of China, what are the standard codes representing the cosmetics sector? (　　)
   A. GB and GB/T　　B. GB and QB/T　　C. QB and GB/T　　D. QB and QB/T

3. According to Table 4, what's the import tariffs of Shampoos in 2019? (　　)
   A. 1%　　　B. 2%　　　C. 3%　　　D. 5%

4. Which of the following expressions is allowed for use in the names of cosmetic products? (　　)
   A. arbitrary expressions
   B. expressions that falsely claim
   C. expressions that explicitly or implicitly indicate the medical effect of a product
   D. concise, easy to understand

5. According to The Instructions for Consumer Goods-General Labelling of Cosmetics, which content of the cosmetic products labelling may NOT be marked? (　　)
   A. ingredient list
   B. shelf life
   C. net content
   D. the "QS" logo

6. According to the 2015 version of the Safety and Technical Standards for Cosmetics, which chapter is not consisted? (　　)
   A. Physical and Chemical Testing Methods
   B. Microbiological Testing Methods
   C. Toxicological Testing Methods
   D. The Standard Chinese Names of All Added Ingredients In The Product

7. Which of the following is NOT the Special-Use Cosmetics in China? (　　)
   A. hair coloring
   B. breast re-shaping
   C. protection from the sun
   D. shampoos

8. What's the first local legislation regarding the safety supervision of cosmetics in China? (　　)
   A. Beijing Cosmetics Safety Regulations
   B. Shanghai Cosmetics Safety Regulations
   C. Guangdong Cosmetics Safety Regulations
   D. Hongkong Cosmetics Safety Regulations

Lesson 16 China's Import and Trade Regulations

Part B Matching the terms of the following regulations.

1. 化妆品卫生监督条例实施细则	1. Safety and Technical Standards for Cosmetics
2. 中华人民共和国标准化法	2. The Detailed Rules for the Implementation of the Regulations on Cosmetics Hygienic Supervision
3. 消费品使用说明·化妆品通用标签	3. Standardisation Law of the People's Republic of China
4. 化妆品命名规定	4. Circular on Matters Related to Cosmetics Production Permit
5. 化妆品安全技术规范	5. Naming Requirements for Cosmetics
6. 关于化妆品生产许可有关事项的公告	6. Administrative Measures on the Inspection, Quarantine and Supervision of Import and Export of Cosmetics
7. 进出口化妆品检验检疫监督管理办法	7. Guangdong Cosmetics Safety Regulations
8. 跨境电子商务零售进口商品清单(2018年版)	8. List of Cross-Border E-Commerce Retail Imports
9. 关于实施特殊用途化妆品行政许可延续承诺制审批有关事宜的公告	9. The Instructions for Consumer Goods - General Labelling of Cosmetics
10. 广东省化妆品安全条例	10. The Announcement of the NMPA on Matters Concerning the Implementation of the Commitment-based Examination and Approval System for the Renewal of Administrative

知识拓展

广东省化妆品安全条例

Task 2 Hygiene License, Record-keeping Certificate & CIQ Labels for Imported Cosmetics in China

任务二 中国进口化妆品卫生许可证、备案证书及 CIQ 标签

China is the cosmetics market with huge growth potential in the future. According to forecasts, the Chinese cosmetics market will grow by more than 10% each year in the near future. Compared to other industries, it offers more business opportunities and possibilities. But did you know that you shall obtain Hygiene License and CIQ Label first before your place cosmetics on Chinese market?

All imported cosmetics can only be sold in the Chinese market after they have obtained Hygiene License or Record-keeping Certificate for imported cosmetics issued by the SFDA (the State Food and Drugs Administration) and CIQ Labels. A cosmetic with CIQ label means it has already passed the examination of China Entry-Exit Inspection and Quarantine Bureau and is allowed to be sold in China. Consumers will check the CIQ label before purchasing imported cosmetics.

On April 1st 2010, the new rules of Hygiene License Application for imported cosmetics were implemented. We will give detailed guidance on how to obtain Hygiene License and CIQ Label in this document according to new rules.

- **Hygiene License or Record-keeping Certificate**

It is a document issued by the SFDA approving the import of a cosmetics product to China.

Hygiene license is issued for specific use cosmetics and record-keeping certificate is issued for ordinary cosmetics.

- **CIQ (China Inspection Quarantine) Labels**

It is a laser label adhered onto the packages of regular import cosmetics which have been approved by China Entry-Exit Inspection and Quarantine Bureau for sale in the Chinese market.

Cosmetics can be divided into two categories and each category requires different procedures.

1. Imported cosmetics for non-specific uses:

For hair, nail (toenail), body care, aroma product, color cosmetics.

Cosmetics for non-specific uses do not require a hygiene license. However, companies shall apply for record-keeping certificates for those products. The first step is to send the sample of a product to a cosmetic testing institution approved by the SFDA for testing.

After receiving application materials along with the testing report, the SFDA will decide whether to issue record-keeping certificate or not within 20 working days.

2. Imported cosmetics for specific uses:

Deodorants, depilation, pimple removal, anti-sunburn, perm, hair dyes, hair grower, body building and breast beauty.

Cosmetics for specific use will need the hygiene license. More tests need to be carried out and it takes longer to obtain such license.

After testing, applicant shall gather all required documents and send them to Safety Approval Committee of the ministry of health for review. It takes 2 to 8 months for the committee to evaluate one product. Usually, the hygiene license will be issued within 1.5 or 2 months after the official evaluation meeting.

- **Application Materials Required (1 original, 13 copies)**

① Application form for hygiene license of imported cosmetic;

② Product ingredient;

③ Effective components, evidence of usage and inspection methods;

④ Manufacturing technique and diagram;

⑤ Product quality standard (in-house standard);

⑥ Testing report from a cosmetic testing institution approved by the SFDA and related materials: Application form for inspection; Notice of acceptance for inspection; Product instruction; Inspection report issued by cosmetic inspection organization approved by the SFDA;

⑦ Product package (sales package & product label);

⑧ Certified document for production and sales in the manufacturing country (region);

⑨ The statement on related problems of 'Mad Cow Disease';

⑩ Apply the power of attorney, if provided by agent;

⑪ Some other documents may be helpful for inspection;

⑫ 3 sealed samples (as the case may be).

If you export cosmetics for specific use to China, you should submit extra materials: effective components, evidence of usage, inspection methods of products and body safety testing reports.

- **Procedures for Application of Hygiene License for Imported Cosmetics**

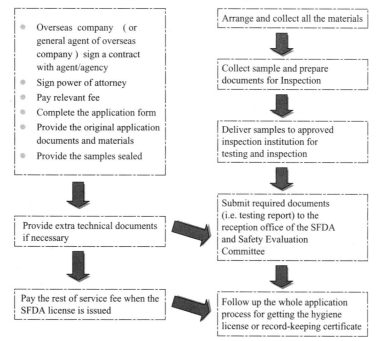

Key Words & Phrases　重点词汇

Hygiene Licence of Imported Cosmetics　进口化妆品卫生许可证
forecast['fɔːkɑːst]　n.（天气、财经等的）预测，预报；预想　v. 预报，预测；预示；预言
Record-keeping Certificate　备案证书
the State Food and Drugs Administration (SFDA)　国家食品药品监督管理局
China Inspection Quarantine (CIQ) Labels　检验检疫标志
China Entry-Exit Inspection and Quarantine Bureau　国家出入境检验检疫局
depilation[ˌdepɪˈleɪʃən]　n. 脱毛
pimple removal　祛斑
perm[pɜːm]　n. 电烫发
Safety Approval Committee of the ministry of health　卫生部化妆品卫生安全性评审委员会
application form　申请表
in-house standard　机构内部标准
mad cow disease　疯牛病
the power of attorney (POA)　授权委托书
sealed[siːld]　adj. 密封的；未知的
inspection institution　检验机构

Practical Activities 实践活动

Part A Reading comprehension.

1. According to the passage, which document is required for specific use cosmetics when imported to China? (　　)
   A. application form   B. hygiene license
   C. record-keeping certificate   D. power of attorney

2. According to the passage, which document is required for non-specific use cosmetics when imported to China? (　　)
   A. application form   B. jygiene license
   C. record-keeping certificate   D. power of attorney

3. How long will the SFDA take to decide whether to issue record-keeping certificate? (　　)
   A. 10 working days   B. 20 working days
   C. 1 month   D. 2 months

4. How long will the Safety Approval Committee of the ministry of health take to evaluate specific use cosmetics? (　　)
   A. 1.5 to 2 months   B. 2 to 8 months
   C. 1.5 to 2 months   D. 2 to 9 months

5. Which label of the imported cosmetics must be labelled before came into Chinese market? (　　)
   A. GB   B. QB
   C. CIQ labels   D. The "QS" logo

6. Which of the following is NOT the imported cosmetics for non-specific uses in China? (　　)
   A. toenail   B. body care
   C. color cosmetics   D. deodorant

7. Which of the following is NOT the imported cosmetics for specific uses in China? (　　)
   A. hair dyes   B. breast beauty
   C. anti-sunburn   D. aroma product

8. Which application material is NOT required when applicant want to obtain the hygiene license of imported cosmetics? (　　)
   A. application form   B. product ingredient
   C. manufacturing technique and diagram   D. CIQ labels

Part B Matching the terms of the following regulations.

```
1. 进口化妆品卫生许可证          1. Record-keeping Certificate
2. 备案证书                      2. Hygiene License of Imported Cosmetics
3. 检验检疫标志                  3. China Entry-Exit Inspection and Quarantine Bureau
4. 国家出入境检验检疫局          4. inspection institution
5. 卫生部化妆品卫生安全性评审委员会  5. China Inspection Quarantine (CIQ)Labels
6. 检验机构                      6. Safety Approval Committee of the ministry of health
7. 国家食品药品监督管理局        7. the State Food and Drugs Administration (SFDA)
```

知识拓展

Unit Five
Cosmetic Safety
化妆品安全

Learning Objectives 学习目标

In this unit you will be able to:
1. Have an overview of cosmetic safety information including safety and technical standards for cosmetics, cosmetic ingredient review, safety assessment and cosmetic product safety report.
2. Learn how to safely use eye cosmetics and how to safely use nail care products.
3. Understand factors considered in safety evaluation and product safety in the marketplace.
4. Study the flavor and extract manufacturers association of the United states.

Lesson 17 Cosmetic & Personal Care Product Safety
化妆品及个人护理产品的安全管理

Lead in 课前导入

Thinking and talking:
1. What the FDA says about cosmetic safety?
2. Do you know the safety and technical standards for cosmetics?

Related words:

U. S. Food and Drug Administration (FDA)	美国食品药品监督管理局
Administration	行政部门
safety and technical standards for cosmetics	化妆品安全技术标准

Task 1 What the FDA Says about Cosmetic Safety
任务一 法律是怎么规定化妆品安全的

People use cosmetics to keep clean and enhance their beauty. These products range from lip-

Lesson 17 Cosmetic & Personal Care Product Safety

stick and nail polish to deodorant, perfume, hairspray, shampoo, shower gel, tattoos, hair adhesives, hair removal products, hair dyes, most soaps, some tooth whiteners, and some cleansing wipes. It's important to use cosmetics products safely. The U.S. Food and Drug Administration (FDA) reminds you to get the facts before using cosmetics products. FDA monitors the safety of cosmetics in several ways. For example, FDA periodically buys cosmetics and analyzes them, especially if we are aware of a potential problem. FDA scientists keep up with the latest research and FDA conducts its own research as well. We also evaluate reports of problems that are sent to us by consumers who have had bad reactions, to watch for trends that will tell us if a particular product may require action on FDA's part. When we look into the safety of a cosmetic product or ingredient on the market, we consider factors such as how it is used and who is likely to use it. This includes whether there are likely to be safety concerns when women use the product during pregnancy. When we identify a safety problem, we let the public know and take action against the product.

The law does not require cosmetics to be approved by FDA before they are sold in stores. However, FDA does monitor consumer reports of adverse events with cosmetic products. Please notify FDA if you experience a rash, redness, burn, or another unexpected reaction after using a cosmetic product. Also, please contact FDA if you notice a problem with the cosmetic product itself, such as a bad smell, color change, or foreign material in the product.

It's important to know that the law does not require cosmetic products or ingredients to have FDA approval before they go on the market. However, cosmetics must be safe when consumers use them according to product labeling, or as the products are customarily used. Companies and individuals who manufacture or market cosmetics are legally responsible for making sure their products are safe. FDA can take action against an unsafe cosmetic that doesn't comply with the law, but first we need reliable information showing that it is unsafe when people use it as intended.

The law treats color additives differently. Color additives must be approved by FDA before they are used in cosmetics or other FDA-regulated products. Some must even be from batches certified in FDA's own labs.

The law also makes a special exception for coaltar hair dyes, which include most permanent, semi-permanent, and temporary hair dyes on the market. FDA cannot take action against a coal-tar hair dye if it has this statement on the label—"Caution: This product contains ingredients which may cause skin irritation on certain individuals and a preliminary test according to accompanying directions should first be made. This product must not be used for dyeing the eyelashes or eyebrows; to do so may cause blindness."—along with instructions for doing a skin test.

Key Words & Phrases 重点词汇

rash[ræʃ] *n.* 皮疹

temporary['temprəri] *adj.* 短暂的；暂时的；临时的 *n.* 临时工；临时雇员

lipstick['lɪpstɪk]　　n. 口红

periodically[ˌpɪərɪ'ɒdɪkəli]　　adv. 定期；周期性

Practical Activities　实践活动

Translation exercise.

1. When we look into the safety of a cosmetic product or ingredient on the market, we consider factors such as how it is used and who is likely to use it.

2. FDA cannot take action against a coal-tar hair dye if it has this statement on the label—"Caution-This product contains ingredients which may cause skin irritation on certain individuals and a preliminary test according to accompanying directions should first be made. This product must not be used for dyeing the eyelashes or eyebrows; to do so may cause blindness."

3. Cosmetics must be labeled properly. For example, they must have any directions for use and any warnings needed to make sure consumers use the product safely.

Task 2　Safety and Technical Standards for Cosmetics

任务二　化妆品安全技术标准

Cosmetics must be labeled properly. For example, they must have any directions for use and any warnings needed to make sure consumers use the product safely.

Also, cosmetics marketed on a retail basis to consumers, such as in stores or person to person, must have a list of ingredients on the label. For cosmetics sold by mail order, including online, this list must be on the label, in a catalog, on a website, enclosed with the shipment, or sent separately when the consumer asks for it. This list lets consumers know if a product contains ingredients they want to avoid.

Key Words & Phrases　重点词汇

keep up with　跟上

in stores　店内

person to person　面对面

be aware of　意识到

Practical Activities　实践活动

Group discussion tasks.

Discuss the following question with your partners

1. What are the safety and technical standards for cosmetics?

For example, FDA periodically buys cosmetics and analyzes them, especially if we are

Lesson 17 Cosmetic & Personal Care Product Safety

aware of a potential problem.

2. Give examples of the safety of cosmetics.

For example, they must have any directions for use and any warnings needed to make sure consumers use the product safely.

3. How to treat the color additives?

Color additives must be approved by FDA before they are used in cosmetics or other FDA-regulated products. Some must even be from batches certified in FDA's own labs.

知识拓展

Lesson 18 Cosmetic Ingredient Review (CIR) 化妆品成分评估

Lead in 课前导入

Thinking and talking:
1. Introduction of cosmetic ingredient review.
2. Learn to check CIR findings.
3. Safety assessment of cosmetic ingredients.

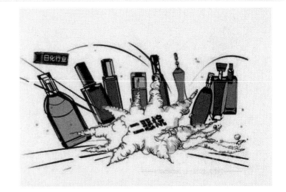

Related words:

ingredient	成分
council	(市、郡等的)政务委员会，地方议会
toiletry	化妆品，化妆用具

Lesson 18　Cosmetic Ingredient Review（CIR）

identification	鉴定；辨认；确认；确定；身份证明
qualification	（通过考试或学习课程取得的）资格
definition	（尤指词典里的词或短语的）释义，解释
assessment	看法；评估

Task 1　Introduction of Cosmetic Ingredient Review（CIR）

任务一　了解化妆品成分评估

The Cosmetic Ingredient Review（CIR）studies individual chemical compounds as they are used in cosmetic products. For 40 years CIR has worked with FDA, the cosmetics industry, and consumers to help keep cosmetics safe. CIR was established in 1976 by the Industry Trade Association (then the Cosmetic, Toiletry, and Fragrance Association, now the Personal Care Products Council), with the support of the U.S. Food and Drug Administration and the Consumer Federation of America. Although funded by the council, CIR and the review process are independent from the council and the cosmetics industry.

The purpose of CIR is to determine those cosmetic ingredients for which there is a reasonable certainty in the judgment of competent scientists that the ingredient is safe under its conditions of use. The Expert Panel may, on its own initiative, or at the request of the Chair of the Steering Committee or FDA, or in response to public comment, assign a special priority for and undertake a review of any ingredient (s) that has been identified as deserving expedited review for use in cosmetics.

Key Words & Phrases　重点词汇

reasonable['ri:znəbl]　*adj*. 公平的；合理的；有理由的

certainty['sɜːtnti]   n. 确实的事；必然的事；确信；确实；确定性
initiative[ɪ'nɪʃətɪv]   n. 倡议；新方案；主动性；积极性
association[əˌsəʊsɪ'eɪʃn]   n. 协会；社团；联盟；联合；合伙

Practical Activities   实践活动

Translation exercise.

1. The purpose of CIR is to determine those cosmetic ingredients for which there is a reasonable certainty in the judgment of competent scientists that the ingredient is safe under its conditions of use.

2. Food and Drug Administration and the Consumer Federation of America. Although funded by the Council, CIR and the review process are independent from the Council and the cosmetics industry.

Task 2   Learn to Check CIR Findings
任务二   学习查找化妆品成分评估文件

CIR findings are listed at a website. CIR findings are classified into four categories as shown in Appendix XX: Safe As Used, Safe with Qualifications, Unsafe, Use Not Supported.

Safe As Used (S)	safe in the present practices of use and concentration
Safe With Qualifications (SQ)	safe for use in cosmetics, with qualifications I-the available data are insufficient to support safety Z-the available data are insufficient to support safety, but the ingredient is not used
Unsafe (U)	the ingredient is unsafe for use in cosmetics
Use Not Supported (UNS)	ingredients for which the data are insufficient and their use in cosmetics is not supported

About the Cosmetic Ingredient Review

The Cosmetic Ingredient Review was established in 1976 by the industry trade association (then the Cosmetic, Toiletry, and Fragrance Association, now the Personal Care Products Council), with the support of the U.S. Food and Drug Administration and the Consumer Federation of America. Although funded by the Council, CIR, the Expert Panel for Cosmetic Ingredient Safety, and the review process are independent from the Council and the cosmetics industry. CIR and the Expert Panel for Cosmetic Ingredient Safety operate under a set of procedures.

General policy and direction are given by a 7-member Steering Committee chaired by the President and CEO of the Council, with a dermatologist representing the American Academy of Dermatology, a toxicologist representing the Society of Toxicology, a consumer representative representing the Consumer Federation of America, an industry scientist (the current chair of the Council's CIR Committee), Chair of the Expert Panel for Cosmetic Ingredient Safety, and the Council's Executive Vice President for Science.

Lesson 18 Cosmetic Ingredient Review (CIR)

Expert Panel for Cosmetic Ingredient Safety

40 Years of CIR

For 40 years the Cosmetic Ingredient Review has worked with FDA, the cosmetics industry, and consumers to help keep cosmetics safe.

Find Ingredient Reviews and Documents

Select a letter of the alphabet to browse ingredients by name or enter an ingredient name in the search field below.

Ingredients of interest may be found on the label of a personal care product and searched using this database. Ingredients are often referred to on the label of a personal care product by the ingredient name as it appears in the International Nomenclature of Cosmetic Ingredients (INCI), known as the INCI Name. CIR reviews ingredients, not products. Furthermore, as a matter of practice, CIR does not usually review fragrances, colors, or flavorings.

CIR performs an extensive search of the world literature as part of its preparation of a safety assessment. The websites and sources that CIR routinely searches for information that could be applicable to the preparation of a safety assessment are identified here. Other sources may also be searched, as appropriate. Typical Search Terms INCI names CAS numbers chemical/technical names additional terms will be used as appropriate Search Engines Pubmed (http://www.ncbi.nlm.nih.gov/pubmed).

Acetyl ethyl tetramethyl tetralin (AETT): In a subchronic toxicity study in rats conducted in 1977, AETT was found to cause serious neurotoxic disorders and discoloration of internal organs. It was also determined to penetrate human skin. The fragrance industry voluntarily discontinued the use of AETT in 1978. Investigate and document any use of AETT in fragrance formulations and finished cosmetic products, usually those claiming to be fragrance free.

6-Methylcoumarin (6-MC): 6-MC, a fragrance ingredient, is a potent photocontact sensitizer which may cause serious skin and systemic disorders in some consumers on contact in the presence of sunlight. Between 1976 & 1978, the FDA received many reports of adverse reactions associated with the use of 6-MC containing suntan preparations. The photocontact allergenicity of 6-MC was subsequently confirmed in clinical studies. In 1978, the FDA asked manufacturers of suntan and sunscreen products to discontinue the use of 6-MC. Two firms voluntarily recalled their 6-MC containing suntan products from the market. Investigate and document any use of 6-MC in the fragrance of sun exposure products.

Musk Ambrette: Musk ambrette, a fragrance ingredient, may cause photocontact sensitization, i.e., allergic reaction of the skin on exposure to musk ambrette and sunlight. Animal studies demonstrated that musk ambrette may cause neurotoxic effects. The International Fragrance Association has recommended that musk ambrette should not be used in products applied to the skin, particularly in products used on skin that is customarily also exposed to sunlight. Investigate and document any use of musk ambrette in the fragrance of sun exposure products.

Nitrosamines: Cosmetics containing as ingredients amines and amino derivatives, particularly di-& triethanolamine (DEA & TEA) may form nitrosamines, if they also contain an ingredient which acts as a nitrosating agent as for example, 2-bromo-2-nitropropane-1,3-diol(Bronopol, Onyxide 500), 5-bromo-5-nitro-1,3-dioxane(Bronidox C) or tris (hydroxymethyl) nitro-methane (Tris Nitro); or if they are contaminated with a nitrosating agent, e.g., sodium nitrite. Amines and their derivatives are mostly present in creams, cream lotions, hair shampoos and cream hair conditioners. The nitrosation may occur during manufacture as well as product storage.

Many nitrosamines have been determined to cause cancer in laboratory animals. They have also been shown to penetrate the skin. Nitrosamine contamination of cosmetics became an issue in early 1977. In a study of 29 cosmetic creams and lotions, N-Nitrosodiethanolamine (NDELA) was determined in 27. The levels of NDELA contamination ranged from less than 10 ppb to 50 ppm. Of the more than 300 cosmetic samples analyzed in 1978, 1979 and early 1980 in FDA laboratories, 7% contained less than 30 ppb NDELA, 26% contained 30 ppb to 2 ppm, and 7% contained between 2 ppm and 150 ppm.

The FDA expressed its concern about the contamination of cosmetics with nitrosamines in a federal register notice dated April 10th, 1979, which stated that cosmetics containing nitrosamines may be considered adulterated and subject to enforcement action. In surveys of cosmetic products conducted in 1991-1992, NDELA was found in 65% of the samples at levels up to 3 ppm.

Investigate whether DEA or TEA containing products contain as ingredients one of the aforementioned nitrosating agents, and report any cosmetic containing these two types of ingredients. When collecting surveillance samples, select such products for chemical analysis.

Dioxane: Cosmetics containing as ingredients ethoxylated surface active agents, i.e., detergents, foaming agents, emulsifiers and certain solvents identifiable by the prefix, word or syllable "PEG" "Polyethylene" "Polyethylene glycol" "Polyoxyethylene" "-eth-" or "-oxynol-" may be contaminated with 1,4-dioxane. It may be removed from ethoxylated compounds by means of vacuum stripping at the end of the polymerization process without an unreasonable increase in raw material cost.

In rodent feeding studies conducted for the National Cancer Institute, 1,4-dioxane was found to produce cancer of the liver and the nasal turbinates. It also caused systemic cancer in a skin painting study. Skin absorption studies demonstrated that dioxane readily penetrates animal and human skin from various types of vehicles. However, it was also determined that most of the dioxane applied to the skin in a vehicle evaporates into the environment and may not be available for skin absorption.

The contamination of ethoxylated surface-active agents with dioxane was first reported in 1978. Many of the raw materials analyzed since then have been found to contain dioxane; some contained as much as, or more than, 100 ppm. In finished cosmetic products containing ethoxylated surface-active agents, the incidence and level of dioxane contamination was significantly lower.

Key Words & Phrases 重点词汇

rodent['rəʊdnt] *n.* 啮齿动物
contamination[kənˌtæmɪ'neɪʃn] *n.* 污染；污秽；(语言的)交感
polymerization[ˌpɒlɪməraɪ'zeɪʃn] *n.* 聚合；多项式
agent['eɪdʒənt] *n.* (企业、政治等的)代理人，经纪人

investigate[ɪnˈvestɪɡeɪt] v. 侦查（某事）；调查（某人）；研究；调查

Practical Activities 实践活动

Translation exercise.

1. CIR Ingredient Status Report, i. e. Safety Assessment, can be searched using a database.
2. CIR Findings are classified into four categories as shown in Appendix XX: Safe As Used, Safe with Qualifications, Unsafe, Use Not Supported.

Task 3 Safety Assessment of Cosmetic Ingredients

任务三 化妆品成分的安全评估报告

CIR ingredient status report, i. e. safety assessment, can be searched using a database (https://www.cir-safety.org/ingredients). Appendix XX show the safety assessment of two cosmetic ingredients, which will be excerpted in Unit XX.

Generally, the report of safety assessment includes:

(1) Chemistry. Definition and structure; Plant identification (to be used in botanical reports); Physical and chemical properties (present measured properties first, computational second); Method of manufacture; Composition (generally, included for botanicals); Impurities; Other sub-sections as appropriate (for example: Natural occurrence; UV absorption; Nitrosation).

(2) Cosmetic and non-cosmetic use.

(3) Toxicokinetic studies, including dermal penetration, penetration enhancement, absorption, distribution, metabolism, and excretion (ADME).

(4) Toxicological studies, including acute toxicity studies, short-term toxicity studies, subchronic toxicity studies, chronic toxicity studies.

(5) Developmental and reproductive toxicity (DART) studies.

(6) Genotoxicity studies.

(7) Carcinogenicity studies.

(8) Anti-carcinogenicity studies.

(9) Other relevant studies, such as comedogenicity, effects on pigmentation, endocrine effects, cytotoxicity, anti-microbial and other endpoints.

(10) Dermal irritation and sensitization studies such as irritation, sensitization and photo-sensitization/phototoxicity.

(11) Ocular irritation studies.

(12) Mucous membrane irritation studies.

(13) Clinical studies such as retrospective and multicenter studies, case reports, adverse event reports.

Lesson 18 Cosmetic Ingredient Review (CIR)

(14) Epidemiological studies.

Key Words & Phrases 重点词汇

website['websaɪt] *n.* 网站

association[ə,səʊsi'eɪʃn] *n.* 协会；社团；联盟

toxicological[ˌtɒksəkə'lɑdʒɪkəl] *n.* 毒理学

toxicity[tɒk'sɪsəti] *n.* 毒性，毒力

ocular['ɑkjələ] *adj.* 眼睛的；视觉的；目击的 *n.* 目镜

relevant['reləvənt] *adj.* 相关的；切题的；中肯的；有重大关系的；有意义的，目的明确的

Practical Activities 实践活动

Part A　Reading comprehension.

1. CIR was established in (　　).
A. 1976　　　　B. 1997　　　　C. 1967　　　　D. 1999

2. What is the purpose of CIR? (　　)

A. To determine those cosmetic ingredients for which there is a reasonable certainty in the judgment of competent scientists

B. To remind you to get the facts before using cosmetics products

C. To flavor manufacturers, flavor users, flavor ingredient suppliers, and others

D. an acronym for the phrase "generally recognized as safe"

3. How to get the CIR Ingredient Status Report? (　　)

A. using a database　B. newspaper　　C. magazine　　D. TV

Part B　Translation exercise.

1. CIR Ingredient Status Report, i.e. Safety Assessment, can be searched using a database.

2. CIR Findings are classified into four categories as shown in Appendix XX: Safe As Used, Safe with Qualifications, Unsafe, Use Not Supported.

知识拓展

Lesson 19 FEMA GRAS Program
FEMA 的公认安全、无毒项目

Lead in 课前导入

Thinking and talking:
1. The flavor and extract manufacturers association of the United States.
2. FEMA GRAS program.
3. Learn to check GRAS notice inventory.

Related words:

flavor	味道	initial	最初的，开始的，首字母
legislator	立法者	status	地位
substance	物质	review	复习，评审

Task 1 The Flavor and Extract Manufacturers Association of the United States

任务一　认识美国香精和提取物制造商协会

The Flavor and Extract Manufacturers Association of the United States (FEMA) is comprised of flavor manufacturers, flavor users, flavor ingredient suppliers, and others with an interest in the U. S. flavor industry. Founded in 1909, it is the national association of the U. S. flavor industry. FEMA works with legislators and regulators to assure that the needs of members and consumers are continuously addressed. FEMA is committed to assuring a substantial supply of safe flavoring substances.

FEMA has developed a program utilizing the generally recognized as safe (GRAS) concept to evaluate the safety of flavoring substances. The independent FEMA Expert Panel determines the GRAS status of flavoring substances resulting in a GRAS list of over 2,800 ingredients for use by the industry. The Expert Panel periodically reviews the status of substances and provides an opportunity for the introduction of new flavoring substances.

Key Words & Phrases 重点词汇

substance['sʌbstəns] *n.* 物质；物品；东西
generally['dʒenrəli] *adv.* 普遍地；广泛地；一般地；通常
continuously[kən'tɪnjʊəsli] *adv.* 连续不断地
manufacturer[ˌmænjʊ'fæktʃərə(r)] *n.* 生产者；制造者；生产商
status['steɪtəs] *n.* 状态；地位；身份；职位

Practical Activities 实践活动

Translation exercise.

1. FEMA has developed an program utilizing the generally recognized as safe (GRAS) concept to evaluate the safety of flavoring substances.

2. The Expert Panel periodically reviews the status of substances and provides an opportunity for the introduction of new flavoring substances.

Task 2 FEMA GRAS Program

任务二 美国香精和提取物制造商协会的公认安全无毒项目

In 1959, FEMA took its initial actions to establish a novel program to assess the safety and "GRAS" status of flavor ingredients as described in the 1958 Food Additives Amendments to the Federal Food, Drug, and Cosmetic Act, the Federal law governing the regulation of flavors and other food ingredients. Since then, the FEMA GRAS program has become the longest-running and most widely recognized industry GRAS assessment program. The FEMA GRAS program began in 1959 with a survey of the flavor industry to identify flavor ingredients then in use and to provide estimates of the amounts of these substances used to manufacture flavors.

"GRAS" is an acronym for the phrase Generally Recognized As Safe. Under sections 201 (s) and 409 of the Federal Food, Drug, and Cosmetic Act, any substance that is intentionally added to food is a food additive, that is subject to premarket review and approval by FDA, unless the substance is generally recognized, among qualified experts, as having been adequately shown to be safe under the conditions of its intended use, or unless the use of the substance is otherwise excepted from the definition of a food additive.

Key Words & Phrases 重点词汇

initial[ɪˈnɪʃl]   adj. 最初的；开始的；第一的   n.（名字的）首字母；（全名的）首字母
acronym[ˈækrənɪm]   n. 首字母缩略词
unless[ənˈles]   conj. 除非；除非在……情况下
otherwise[ˈʌðəwaɪz]   adv. 否则；不然；除此以外

Practical Activities 实践活动

Group discussion tasks.

How to evaluate the safety of flavoring substances?

FEMA has developed an program utilizing the generally recognized as safe (GRAS) concept.

Give an example of the inventory of GRAS. (For example: GRAS Notice of Patchouly Oil)

Task 3   Learn to Check GRAS Notice Inventory

任务三　学习查找公认安全无毒清单

The inventory of GRAS is listed at a website. Here are two examples (table 1 and table 2).

Table 1　GRAS Notice of Patchouly Oil

CAS Reg. No. (or other ID)	8014-9-3
Substance	Patchouly, Oil (Pogostemon Spp.)
Other Names	◆ Patchouli Oil
	◆ Patchouly Oil
	◆ Oils, Patchouli
	◆ Pogostemon Oil
	◆ Pogostemon Cablin Oil

续表

CAS Reg. No. (or other ID)	8014-9-3
Used for (Technical Effect)	Flavoring Agent or Adjuvant
Food additive and GRAS regulations (21 CFR Parts 170-186)	172.51
FEMA No.	2838
FEMA GRAS Publication No(s).	3

Table 2　GRAS Notice of Nerol

CAS Reg. No. (or other ID)	106-25-2
Substance	Nerol
Other Names	◆ Nerol
	◆ Neryl Alcohol
	◆ 3,7-Dimethyl-2,6-Octadien-1-Ol,Cis-
	◆ 2,6-Octadien-1-Ol,3,7-Dimethyl-,(Z)-
	◆ 3,7-Dimethyl-2,6-Octadien-1-Ol,(Z)-
	◆ Geraniol,Cis-
	◆ Geraniol,(Z)-
	◆ Nerol,Beta-
	◆ Nsc-46105
	◆ Beta-Nerol
Used for (Technical Effect)	Flavoring Agent or Adjuvant
Food additive and GRAS regulations (21 CFR Parts 170-186)	172.515
FEMA No.	2770
FEMA GRAS Publication No(s).	3
JECFA Flavor Number	1224

CAS Reg. No. (or other ID): Chemical Abstract Service (CAS) Registry Number for the substance or a numerical code assigned by CFSAN to those substances that do not have a CAS Registry Number (977nnn-nn-n series).

21 CFR: Title 21 of the Code of Federal Regulations.

FEMA No.: The trade association, Flavor and Extract Manufacturers Association (FEMA), has established expert panels to evaluate and make independent determinations on the GRAS status of flavoring substances. The FEMA number is provided here as a reference to FEMA's GRAS assessments.

JECFA: The Joint Expert Committee on Food Additives (JECFA) is an international expert scientific committee that is administered jointly by the Food and Agriculture Organization of the United Nations (FAO) and the World Health Organization (WHO).

Key Words & Phrases 重点词汇

assessment[əˈsesmənt]　n. 评估
assure[əˈʃʊə;əˈʃɔː]　vt. 向……保证；使……确信
acronym[ˈækrənɪm]　n. 首字母缩略词
premarket[priːˈmɑːkɪt]　adj. 上市（销售）前的
be comprised of　由……组成
since then　从那时起
in use　依次，轮流地

Practical Activities 实践活动

Reading comprehension.

1. What is not the right description of "GRAS"? (　　)
A. "GRAS" is an acronym for the phrase Generally Recognized As Safe
B. It's under sections 201（s）and 409 of the Federal Food，Drug
C. That is subject to premarket review and approval by FDA
D. The FEMA GRAS program began in 1969

2. which is not the composition of FEMA? (　　)
A. lavor manufacturers　　　B. flavor users
C. flavor ingredient suppliers　　D. drug

知识拓展

国内化妆品安全规范及相关会议

Lesson 20 Shelf Life and Expiration Dating of Cosmetics

化妆品的货架寿命和保质期

Lead in 课前导入

Thinking and talking:
1. Factors affect shelf life.
2. How to keep your cosmetics safe.

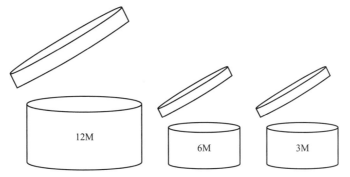

Related words:

| expiration | 满期；截止；呼气 | label | 标签 |

| substance | 物质 | status | 地位 |
| substantiate | 证实；实体化 | review | 复习，评审 |

Task 1 Factors Affect Shelf Life

任务一 影响货架寿命的因素

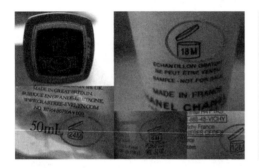

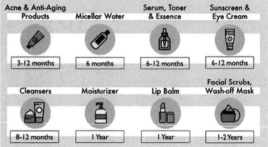

The shelf life, or expiration date, of a cosmetic or personal care product is the period during which the manufacturer has determined a product to be best suited for use.

There are no regulations or requirements under current U. S. laws that require cosmetic manufacturers to print expiration dates on the labels of cosmetic products. However, manufacturers have the responsibility to determine shelf life for products as part of their responsibility to substantiate product safety.

Expiration dates are required for over-the-counter (OTC) drugs. However for OTC drugs without dosage limitations (e. g., antiperspirants, antidandruff shampoos, toothpastes, sunscreens, etc.), OTC drug regulations (21 CFR 211.137) do not require an expiration date provided that these products have demonstrated at least 3 years of stability.

In Europe, cosmetic products with a lifespan longer than 30 months must show a "period after opening" (POA) time. That is, the time in months when the product will remain in good

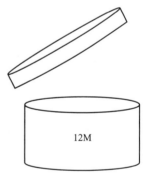

condition after the consumer has used the product for the first time. A symbol of an open cream jar is usually used instead of words and the time in months can be inside the symbol or alongside it. Although this symbol is frequently present on some U. S. cosmetics products, it is not required.

There are no U. S. laws or regulations that require cosmetics to have specific shelf lives or have expiration dates on their labels. However, manufacturers are responsible for making sure their products are safe. FDA considers determining a product's shelf life to be part of the manufacturer's responsibility.

Not all "personal care products" are regulated as cosmetics. Some, such as sunscreen products and acne treatments, are drugs under the law. Some, such as makeup and moisturizers that are also sunscreens, with "SPF" labeling, are regulated as both cosmetics and drugs. Drugs, including those that are both drugs and cosmetics, must be tested for stability (see the regulations at 21 CFR 211) and are required to have expiration dates printed on the labels. Manufacturers must make sure their drug products are safe and effective until their expiration dates.

Key Words & Phrases   重点词汇

shelf[ʃelf]   n.（固定在墙上的或橱柜、书架等的）架子，搁板；（悬崖上或海底）突出的岩石；陆架；陆棚

sunscreen['sʌnskriːn]   n. 防晒霜；防晒油

lifespan[ˌlaɪfspæn]   n. 寿命；可持续年限；有效期

expiration[ˌekspəˈreɪʃn]   n. 告终；期满；截止

Practical Activities   实践活动

Translation exercise.

1. The shelf life, or expiration date, of a cosmetic or personal care product is the period during which the manufacturer has determined a product to be best suited for use.
2. There are no regulations or requirements under current U.S. laws that require cosmetic manufacturers to print expiration dates on the labels of cosmetic products.
3. In Europe, cosmetic products with a lifespan longer than 30 months must show a "period after opening" (POA) time.
4. There are no U.S. laws or regulations that require cosmetics to have specific shelf lives or have expiration dates on their labels.

Task 2   How to Keep Your Cosmetics Safe

任务二   如何保持化妆品安全

How long you can use a cosmetic safely also depends on you. Here are tips to help keep your cosmetics safe.

- If mascara becomes dry, throw it away. Do not add water or, even worse, saliva to moisten it, because that will introduce bacteria into the product. If you have an eye infection, talk with your health care provider, stop using all eye-area cosmetics, and throw away those you were using when the infection occurred.
- Don't share makeup. You may be sharing an infection. "Testers" at cosmetic counters in stores are even more likely to become contaminated than the same products in your home. If you feel you must test a cosmetic before you buy it, apply it with a new, unused applicator, such as a fresh cotton swab.
- Keep containers and applicators clean.
- Store cosmetics properly. For example, don't leave them where they are exposed to

heat, such as in a hot car. Heat can make preservatives break down and cause bacteria and fungi to grow faster.
- Be wary of products offered for sale in flea markets or re-sold over the Internet. Some may be past their shelf life, already used, diluted, or tampered with in other ways. They may even be counterfeit, "fake" versions of the product you think you're buying.

There are best practices for using cosmetics and personal care products.
- Read the instructions carefully and take note of any warnings for use.
- Tightly close lids on products when they are not in use.
- Use products within the lifespan indicated by the Period After Opening symbol or best before date.
- Avoid storing products in direct sunlight or near sources of heat; choose cool (not freezing) areas where possible.
- Never dilute products (e.g., mixing water in mascara).
- Apply products with clean hands or an applicator.
- Wash applicators thoroughly with soap, detergent or a mild shampoo then allow to dry completely before use.
- Avoid sharing cosmetics and personal care products with another person.
- Apply with new, unused applicator when testing at department store or cosmetic counter.
- When in doubt, throw it out!

Key Words & Phrases　重点词汇

dosage[ˈdəʊsɪdʒ]　*n.* 剂量，用量
counterfeit[ˈkaʊntəfɪt]　*n.* 赝品；伪造品　*adj.* 假冒的；假装的　*v.* 仿造；伪装；假装
alongside[əlɒŋˈsaɪd]　*adv.* 在……的侧面；在……旁边；与……并排　*prep.* 在……旁边；横靠；傍着
infection[ɪnˈfekʃ(ə)n]　*n.* (医)传染，感染；传染病，染毒物；影响
lifespan[ˈlaɪfspæn]　*n.* 寿命；使用期限

Practical Activities　实践活动

Part A　Translation exercise.

1. The shelf life, or expiration date, of a cosmetic or personal care product is the period during which the manufacturer has determined a product to be best suited for use.
2. There are no regulations or requirements under current U.S. laws that require cosmetic manufacturers to print expiration dates on the labels of cosmetic products.
3. How long you can use a cosmetic safely depends on you.

Part B　Group discussion tasks.

1. Give tips to help keep your cosmetics safe.

For example: Be wary of products offered for sale in flea markets or re-sold over the Internet.

Lesson 20 Shelf Life and Expiration Dating of Cosmetics

2. Do you know why the cosmetic products with a lifespan longer than 30 months must show a "period after opening" (POA) time.

That is, the time in months when the product will remain in good condition after the consumer has used the product for the first time.

知识拓展

Lesson 21 How to Safely Use Cosmetics
如何安全使用化妆品

Lead in 课前导入

Thinking and talking:
1. Are tattoo inks safe?
2. How to safely use eye cosmetics?
3. How to safely use nail care products?

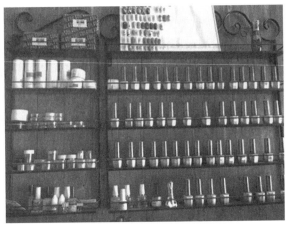

Related words:

tattoo	纹身,(在皮肤上)刺图案
infection	(医)传染,感染;传染病,染毒物;影响
pigment	色素;颜料把……加颜色;变色
adverse	不利的;有害的;逆的;相反的
concealer	隐藏;隐瞒;掩盖
extension	伸展,扩大;延长,延期;牵引;电话分机

Task 1 Are Tattoo Inks Safe?

任务一 如何安全使用纹身墨水

Tattoos are more popular than ever. According to a 2015 Harris Poll, about 3 in 10 (or 29%) people surveyed have at least one tattoo. The U.S. Food and Drug Administration (FDA) is also seeing reports of people developing infections from contaminated tattoo inks, as well as adverse reactions to the inks themselves.

Over the years, the FDA has received hundreds of adverse event reports involving tattoos: 363 from 2004-2016.

Before you get a tattoo, consider these key questions, answered by Dr. Linda Katz, M.D., M.P.H., the director of FDA's Office of Cosmetics and Colors.

1. Should I be concerned about unsafe practices, or the tattoo ink itself?

Both. While you can get serious infections from unhygienic practices and equipment that isn't sterile, infections can also result from ink that was contaminated with bacteria or mold. Using non-sterile water to dilute the pigments (ingredients that add color) is a common culprit, although not the only one.

There's no sure-fire way to tell if the ink is safe. An ink can be contaminated even if the container is sealed or the label says the product is sterile.

2. What is in tattoo ink?

Published research has reported that some inks contain pigments used in printer toner or in car paint. FDA has not approved any pigments for injection into the skin for cosmetic purposes.

FDA reviews reports of adverse reactions or infections from consumers and healthcare providers. We may learn about outbreaks from the state authorities who oversee tattoo parlors.

3. What kinds of reactions may happen after getting a tattoo?

You might notice a rash—redness or bumps—in the area of your tattoo, and you could develop a fever.

More aggressive infections may cause high fever, shaking, chills, and sweats. Treating such infections might require a variety of antibiotics—possibly for months—or even hospitalization and/or surgery. A rash may also mean you're having an allergic reaction. And because the inks are permanent, the reaction may persist.

Key Words & Phrases 重点词汇

antibiotics[ˌæntɪbaɪˈɒtɪks] n. 抗生素（如青霉素）

pigment[ˈpɪgmənt] n. 色素；颜料

Practical Activities 实践活动

Reading comprehension.

1. What is in tattoo ink? (   )
   A. some inks contain pigments   B. saliva or water
   C. curing lamps   D. formaldehyde
2. For eye cosmetics, the CFSAN offer consumers the following advice except (   ).
   A. Keep everything clean   B. Don't dye eyelashes and eyebrows
   C. Exposure to UV radiation   D. Avoid using eye cosmetics if you have an eye infection
3. Which description is wrong? (   )
   A. Manicures and pedicures can be pretty
   B. Nail curing lamps usually come with instructions for exposure time
   C. You can use old eye cosmetics
   D. Nail salon practices are regulated by the states

Task 2 How to Safely Use Eye Cosmetics

任务二 如何安全使用眼妆产品

Lesson 21 How to Safely Use Cosmetics

The Food and Drug Administration (FDA) regulates all cosmetics marketed in the United States, including mascara, eye shadows, eye liner, concealers, and eyebrow pencils. Safety experts within the Office of Cosmetics and Colors in FDA's Center for Food Safety and Applied Nutrition (CFSAN) offer consumers the following advice.

Keep everything clean. Dangerous bacteria or fungi can grow in some cosmetic products, as well as their containers. Cleanliness can help prevent eye infections.

Always wash your hands before applying eye cosmetics, and be sure that any instrument you place near your eyes is clean. Be especially careful not to contaminate cosmetics by introducing microorganisms. For example, don't lay an eyelash wand on a countertop where it can pick up bacteria. Keep containers clean, since these may also be a source of contamination.

Don't moisten cosmetic products. Don't add saliva or water to moisten eye cosmetics. Doing so can introduce bacteria. Problems can arise if you overpower a product's preservative capability.

Don't share or swap. People can be harmed by others' germs when they share eye makeup. Keep this in mind when you come across "testers" at retail stores. If you do sample cosmetics at a store, be sure to use single-use applicators, such as clean cotton swabs.

Don't apply or remove eye makeup in a moving vehicle. Any bump or sudden stop can cause injury to your eye with a mascara wand or other applicator.

Check ingredients, including color additives. As with any cosmetic product sold to consumers, eye cosmetics are required to have an ingredient declaration on the label. If they don't, they are considered misbranded and illegal.

In the United States, the use of color additives is strictly regulated. Some color additives approved for cosmetic use, in general, are not approved for areas near the eyes.

If the product is properly labeled, you can check to see whether the color additives declared on the label are in FDA's List of Color Additives Permitted for Use in Cosmetics.

Use only cosmetics intended for the eyes on the eyes. Don't use a lip liner as an eye liner, for example. You may expose eyes either to contamination from your mouth or to color additives that are not approved for use near the eyes.

Say "no" to kohl! Also known as al-kahl, kajal, or surma, kohl is used in some parts of the world for enhancing the appearance of the eyes. But kohl is unapproved for cosmetic use in the United States.

Kohl contains salts of heavy metals such as antimony and lead. Reports have linked the use of kohl to lead poisoning in children.

Some eye cosmetics may be labeled with the word "kohl" only to indicate the shade, not because they contain true kohl.

A product's "ingredient statement" should not list kohl—this is not an FDA-approved color additive. Check the ingredient statement to make sure that kohl is not present.

Don't dye eyelashes and eyebrows. No color additives are approved by FDA for permanent dyeing or tinting of eyelashes and eyebrows. Permanent eyelash and eyebrow tints and dyes have been known to cause serious eye injuries.

Use care with false eyelashes or extensions. False eyelashes and extensions, as well as their adhesives, must meet the safety and labeling requirements for cosmetics. Since the eyelids are delicate, an allergic reaction, irritation, or injury in the eye area can occur. Check the ingredients to make sure you are not allergic to the adhesives.

Don't use eye cosmetics that cause irritation. Stop using a product immediately if irritation occurs. See a doctor if irritation persists.

Avoid using eye cosmetics if you have an eye infection. Discard any eye cosmetics you were using when you got the infection. Also, don't use eye cosmetics if the skin around the eye is inflamed.

Don't use old eye cosmetics. Manufacturers usually recommend discarding mascara two to four months after purchase. Discard dried-up mascara.

Don't store cosmetics at temperatures above 85°F. Preservatives that keep bacteria or fungi from growing can lose their effectiveness, for example, in cosmetics kept for long periods in hot cars.

Key Words & Phrases　重点词汇

irritation[ˌɪrɪ'teɪʃn]　n. 生气，气恼
vehicle['viːəkl]　n. 交通工具；车辆
moisten['mɔɪsn]　v. (使)变得潮湿，变得湿润

Practical Activities　实践活动

Translation exercise.

1. Always wash your hands before applying eye cosmetics, and be sure that any instrument you place near your eyes is clean.

2. Preservatives that keep bacteria or fungi from growing can lose their effectiveness, for example, in cosmetics kept for long periods in hot cars.

Task 3　How to Safely Use Nail Care Products

任务三　如何安全使用美甲产品

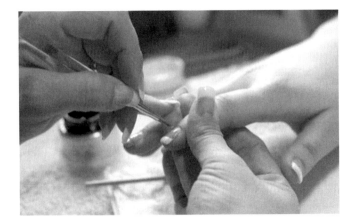

Lesson 21 How to Safely Use Cosmetics

Manicures and pedicures can be pretty. The cosmetic products used, such as nail polishes and nail polish removers, also must be safe—and are regulated by the U. S. Food and Drug Administration.

The FDA also regulates devices used to dry (or "cure") artificial nails or gel nail polish as electronic products because they emit radiation.

You can do your part to stay safe (and look polished, too) by following all labeled directions and paying attention to any warning statements listed on these products.

Cosmetic Nail Care Products: Ingredients and Warnings

Cosmetic ingredients (except most color additives) and products, including nail products, do not need FDA approval before they go on the market.

But these products are required to be safe when used as intended. (Note that nail products intended to treat medical problems are classified as drugs and do require FDA approval.)

Cosmetic nail care products also must include any instructions or warnings needed to use them safely. For example:

Some nail products can catch fire easily so you should not expose them to flames (such as from a lit cigarette) or heat sources (such as a curling iron).

Some can injure your eyes, so you should avoid this exposure.

Some should only be used in areas with good air circulation (ventilation).

Some ingredients can be harmful if swallowed, so these products should never be consumed by any person or pet.

Also know that retail cosmetics such as those sold in stores or online must list ingredients in the order of decreasing amounts. If you're concerned about certain ingredients, you can check the label and avoid using products with those ingredients.

For example, some nail hardeners and nail polishes may contain formaldehyde, which can cause skin irritation or an allergic reaction. And acrylics, used in some artificial nails and sometimes in nail polishes, can cause allergic reactions. (To learn more about ingredients, visit the FDA's nail care products webpage.)

The bottom line? Read the labels of cosmetic products and follow all instructions. And if you go to a salon for a manicure or pedicure, make sure the space has good ventilation.

Note: Nail salon practices are regulated by the states, and not the FDA. If you're a nail salon owner or employee, you can find information on maintaining safe salons on the webpage of the U. S. Department of Labor's Occupational and Health Safety Administration.

If you have questions about whether certain nail products are right for you, talk to your health care provider.

About Nail Drying and Curing Lamps—and UV Exposure

Ultraviolet (UV) nail curing lamps are table-top size units used to dry or "cure" acrylic or gel nails and gel nail polish. These devices are used in salons and sold online. They feature lamps or LEDs that emit UV (ultraviolet) radiation. (Nail curing lamps are different than sunlamps, which are sometimes called "tanning beds." You can learn more about the risks of sunlamps on the FDA's website.)

Exposure to UV radiation can cause damage to your skin, especially if you're exposed over time. For example, it can lead to premature wrinkles, age spots, and even skin cancer.

But the FDA views nail curing lamps as low risk when used as directed by the label. For example, a 2013 published study *External Link Disclaimer* indicated that—even for the worst case lamp that was evaluated—30 minutes of daily exposure to this lamp was below the occupational exposure limits for UV radiation. (Note that these limits only apply to normal, healthy people and not to people who may have a condition that makes them extra sensitive to UV radiation.)

To date, the FDA has not received any reports of burns or skin cancer attributed to these lamps.

That said, if you're concerned about potential risks from UV exposure, you can avoid using these lamps.

You may particularly want to avoid these lamps if you're using certain medications or supplements that make you more sensitive to UV rays. These medications include some antibiotics, oral contraceptives, and estrogens—and supplements can include St. John's Wort. See an extended list of medications that can cause sun sensitivity on the FDA's website.

Also remove cosmetics, fragrances, and skin care products (except sunscreen!) before using these lamps, as some of these products can make you more sensitive to UV rays.

If you have questions about using nail drying or curing lamps, consult a health care professional.

And if you do choose to use these devices, you can reduce UV exposure by:

Wearing UV-absorbing gloves that expose only your nails.

Wearing a broad-spectrum sunscreen with an SPF of 15 or higher. (Since nail treatments can include exposure to water, follow the sunscreen's labeled directions for use in these situations.)

Finally, nail curing lamps usually come with instructions for exposure time. The shorter your exposure, the less risky the exposure, in general. So always follow labeled directions when available. In general, you should not use these devices for more than 10 minutes per hand, per session.

How to Report Problems with Nail Care Products

If you ever have a bad reaction to a cosmetic nail product or nail curing lamp, please consult your health care provider and then tell the FDA.

You can call an FDA Consumer Complaint Coordinator (phone numbers for your area are online) or report the problem via MedWatch, the FDA Safety Information and Adverse Event Reporting program.

Key Words & Phrases   重点词汇

aggressive[ə'gresɪv]　　*adj.* 侵略的；进攻性的；好斗的；有进取心的
sunscreen['sʌnskriːn]　　*n.* (防晒油中的)遮光剂；防晒霜
formaldehyde[fɔː'mældɪhaɪd]　　*n.* 甲醛

Lesson 21 How to Safely Use Cosmetics

spectrum['spektrəm] *n*. 系列；幅度；范围；光谱

Practical Activities 实践活动

Group discussion tasks.

1. How to safely use tattoo?

For example：Before you get a tattoo, consider these key questions, answered by Dr. Linda Katz, M. D., M. P. H., the director of FDA's Office of Cosmetics and Colors.

2. How to safely use eye cosmetics?

For example：Keep everything clean. Dangerous bacteria or fungi can grow in some cosmetic products, as well as their containers. Cleanliness can help prevent eye infections.

3. How to safely use nail care products?

For example：Cosmetic ingredients (except most color additives) and products, including nail products, do not need FDA approval before they go on the market.

知识拓展

Lesson 22 Science Behind Cosmetic Safety
化妆品安全背后的科学

Lead in 课前导入

Thinking and talking:
1. Factors considered in safety evaluation.
2. Product safety in the marketplace.
3. Cosmetic product safety report (CPSR).

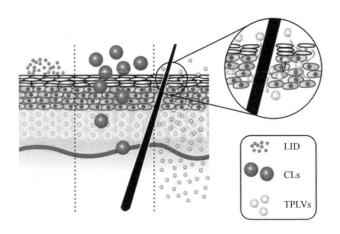

Related words:

reassurance	再保证；再安慰	photoirritation	光刺激
panel	仪表板；嵌板	confirmatory	证实的；确实的
correspondence	通信；信件；相符，相似	exposure	暴露；曝光；揭露；陈列

Task 1　Factors Considered in Safety Evaluation

任务一　评估化妆品安全应考虑的因素

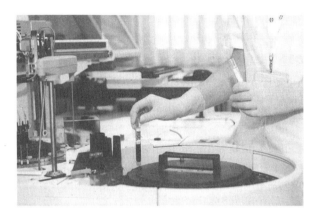

Cosmetics and personal care products companies utilize a multi-tiered scientific approach to extensively evaluate the safety of cosmetic products and ingredients. Key components of this strict and methodical safety process include:

Reviews of Latest, Up-To-Date Safety Research

As part of the scientific cosmetic product evaluation process, numerous scientific research sources are examined for the latest, most up-to-date ingredient safety information. These sources include extensive scientific reviews by the Cosmetic Ingredient Review, an expert panel of independent scientists, as well as reviews of cosmetic ingredients by other sources; information and data from cosmetic ingredient suppliers; published data in the scientific literature; and government sources, including the U.S. Food and Drug Administration, the National Toxicology Program, the National Cancer Institute, and other government databases.

Determinations of Possible Ingredient Toxicology

Safety evaluations take into consideration a number of key factors, including cosmetic ingredient function and use concentration; degree of chemical purity and stability; and potential for cosmetic ingredients to be absorbed through the skin and/or mucous membranes, or via oral ingestion or inhalation. Assessment for potential adverse effects includes: evaluation of exposure to cosmetic ingredients for short, intermediate or long periods of time (acute, subchronic, and chronic systemic toxicity); skin irritation; skin allergy; photoirritation (irritation caused after exposure to sunlight); photoallergy (sensitization caused after exposure to sunlight); and determination of the potential for ingredients to adversely affect the body's genetic material (genotoxicity), cause cancer (carcinogenicity), or negatively effect reproduction and fetal development.

Evaluation & Testing of Human Health Impacts

Following a thorough review of each ingredient in a cosmetic formulation, additional safety data on the finished product are reviewed. These may include cell culture (in vitro) and clinical (human) tests conducted on the final product and on products similar in composition to the product being evaluated. The potential for ingredient interactions within the product leading to unexpected adverse effects also is evaluated. Confirmatory testing of cosmetic product compatibility and acceptability on human volunteers (clinical testing) is often undertaken with informed consent and with the appropriate safeguards to detect any undesirable effects that could occur.

Examination of Cumulative Exposure to the Human Body

The average consumer uses multiple cosmetic and personal care products each day. Therefore, the assessment of cumulative exposure to product ingredients from multiple sources is an important component in the overall assessment of product safety. It is also important to take into account inadvertent (secondary) exposures such as inhalation with hair spray use or ingestion from lipsticks, etc.

Factors to be considered in determining exposure levels for personal care products and ingredients include product type; the amount used per application; frequency of application; site of body contact; duration of product contact; concentration of individual ingredients in the final product; use by sensitive subpopulations such as infants, the elderly, and pregnant women; method of application; external factors such as sunlight exposure and variation in use related to weather, local or temporal habits, trends and cultural considerations; and possible conditions of foreseeable misuse.

Testing & Evaluation Performed by Scientists Trained in Product Safety

To ensure the reliability of testing, cosmetic product safety studies are designed and monitored, and the results are interpreted and evaluated by scientists who are specially trained and experienced in toxicology and safety evaluation. These scientists have a fundamental understanding of cosmetic, personal care and fragrance products and of the tests being used.

Key Words & Phrases 重点词汇

external[ɪkˈstɜːnl] *adj.* 外部的；外面的；外界的　*n.* 外部；外面；外观

exposure[ɪkˈspəʊʒə(r)] *n.* 面临，遭受（危险或不快）；揭露

reliability[rɪˌlaɪəˈbɪlɪti] *n.* 可靠性

Practical Activities 实践活动

Translation exercise.

1. Following a thorough review of each ingredient in a cosmetic formulation, additional safety data on the finished product are reviewed.

2. These scientists have a fundamental understanding of cosmetic, personal care and fragrance products and of the tests being used.

Task 2 Product Safety in the Marketplace

任务二 市场上产品的安全

Safety Monitoring Continues Once Product is on Market

Safety reassurance does not end once a product is placed in the marketplace. Once a product is launched, companies engage in ongoing, active monitoring of consumer experience to confirm product safety. Cosmetic and personal care product manufacturers have established post-market surveillance processes for the identification of potential safety issues related to their products. Such systems help to identify consumer use patterns, such as alternate uses or product combinations that may contribute to adverse events. These processes include regular surveys of consumer contacts received by a marketer or manufacturer either through toll-free 1-800 numbers on packages or direct correspondence. Trend analyses of contact data, including evaluations of frequency and severity of adverse events, as well as comparison of these trends with historical information for other comparable products, represent valuable mechanisms for identification of safety-related concerns. Although adverse reactions that are both serious and unexpected are extremely rare for cosmetic and personal care products, manufacturers must report knowledge of any instances of serious product reactions to the Federal Food and Drug Administration.

The Scientists Behind Product Safety

Scientists that are specially trained in many different disciplines are involved in the design, development, and manufacture of cosmetic and personal care products, including chemistry and biochemistry, microbiology, molecular modeling, engineering, formulation science, technical packaging, and toxicology. While the activity of each of these scientists contributes to the safety profile of cosmetic and personal care products, it is the specific role of the toxicologist to design and interpret the tests that assess the safety of cosmetic and personal care products and their ingredients.

Some of the scientists involved in the product development and safety process, and their roles, include:

toxicologists, who evaluate ingredients and finished products to establish their safety for consumers during usage;

microbiologists, who establish product preservation requirements and monitor manufacturing to ensure finished product integrity (For more information on Preservatives);

analytical chemists, who perform chemical evaluations of ingredients and finished products, to determine purity and other properties;

formulators, who develop new and improved products and set product standards and specifications;

science information specialists, who retrieve information, including scientific articles and patents, relevant to product development and safety;

manufacturing engineers, who develop manufacturing procedures and oversee production of the final product;

technical packaging specialists, who create improved ways of applying, dispensing and packaging products;

quality assurance professionals, who test components and products to meet ingredient and finished product specifications;

regulatory specialists, who ensure that a product meets all labeling requirements and is in compliance with all governmental regulations;

consumer affairs professionals, who conduct consumer research and testing prior to product marketing and track issues that may arise following product marketing.

Key Words & Phrases 重点词汇

affairs[əˈfeə(r)] n. 公共事务；政治事务；事件；事情
compliance[kəmˈplaɪəns] n. 服从；顺从；遵从
track[træk] n. 小道，小径；足迹，踪迹；车辙；轨道 v. 跟踪；追踪

Practical Activities 实践活动

Group discussion tasks.

1. Discuss the key components of this strict and methodical safety process of cosmetic products and ingredients.

For example: Reviews of Latest, Up-To-Date Safety Research. Determinations of Possible Ingredient Toxicology. and so on.

2. Discuss the scientists behind product safety.

Scientists that are specially trained in many different disciplines are involved in the design, development, and manufacture of cosmetic and personal care products, including chemistry and biochemistry, microbiology, molecular modeling, engineering, formulation science, technical packaging, and toxicology.

Task 3　Cosmetic Product Safety Report (CPSR)

任务三　了解化妆品产品安全评估报告

A cosmetic safety assessment or Cosmetic Product Safety Report (CPSR) is a product report which ensures that a cosmetic product is safe for normal conditions of use.

It must be undertaken by a suitably qualified person with a thorough understanding of the EU Cosmetic Regulations.

The safety assessor will consider:

the toxicology of every ingredient submitted for assessment;

the chemical makeup of each ingredient;

the areas on the body where the finished product will be applied;

the type/s of individual who will be using the product (Products intended for those under

Lesson 22 Science Behind Cosmetic Safety

the age of 3 require special reports);

The quantity of each ingredient the intended user will be exposed to.

Nail polish is considered a cosmetic product and therefore nail polish makers in the UK no matter the size of their business must have relevant CPSRs to legally sell their products.

The Micamoma CPSR will cover all Micamoma products at the point of sale.

Prepare a Cosmetic Product Safety Report (CPSR) for your products with SGS.

Cosmetic Products Regulation EC 1223/2009 requires a detailed safety assessment-the CPSR-before products can be marketed within the EU.

Our cosmetic safety assessors work with you to collate and present the details required for the CPSR. Following introduction of the cosmetic product regulation, the CPSR comprises two parts, cosmetic product safety information and a cosmetic product safety assessment.

Requirements for Consumer Products

Labelling and Certification

Industry Canada

Consumer Packaging and Labelling Act regarding bilingual labelling, deceptive packaging and net quantity declaration

Textile Labelling Act & the Textile Labelling and Advertising Regulations

Provincial/Territorial Requirements

Upholstered and Stuffed Articles (Quebec, Ontario and Manitoba)

Mandatory certification of Electrical/Gas Products

New Legislation

Canada Consumer Product Safety Act

New legislation currently going through Parliament (not yet law)

Will replace Part I of the Hazardous Products Act

To address and prevent dangers to human health or safety posed by consumer products (including their components, parts and accessories)

Danger to health and safety: existing or potential hazards during normal/foreseeable use that could lead to death or adverse effect on health (including both acute and chronic health effects)

Prohibitions on the manufacture, importation, sale or advertisement of consumer products

Prohibited as per the schedule

Non-compliant with regulations

Deemed to be a danger to human health or safety ("General Prohibition")

Recalled or the subject of corrective measures that have not been carried out

The subject of false, misleading or deceptive safety claims including false certification marks

> Includes product/hazard-specific regulations
> Ability to require tests to verify compliance
> Record-keeping to allow traceability in the event of a recall
> Authority to require documents/information at importation
> Mandatory reporting by industry on incidents
> Ability to order recall and corrective action
> Increased fines and penalties

Key Words & Phrases　重点词汇

microbiologist [ˌmaɪkrobaɪˈɑlədʒɪst]　*n.* 微生物学家
polish [ˈpɒlɪʃ]　*n.* 光泽；上光剂；优雅；精良
Cosmetic [kɒzˈmetɪk]　*adj.* 美容的；化妆用的　*n.* 化妆品；装饰品
be undertaken by　由……承担
approach to　接近
prior to　在……之前
contribute to　有助于

Practical Activities　实践活动

Part A　Translation exercise.
1. Cosmetics and personal care products companies utilize a multi-tiered scientific approach to extensively evaluate the safety of cosmetic products and ingredients.
2. Following a thorough review of each ingredient in a cosmetic formulation, additional safety data on the finished product are reviewed.
3. A cosmetic safety assessment or Cosmetic Product Safety Report (CPSR) is a product report which ensures that a cosmetic product is safe for normal conditions of use.

Part B　Group discussion tasks.
1. Discuss the key components of this strict and methodical safety process of cosmetic products and ingredients.
For example: Reviews of latest, up-to-date safety research, determinations of possible ingredient toxicology, and so on.
2. Discuss the scientists behind product safety.
Scientists that are specially trained in many different disciplines are involved in the design, development, and manufacture of cosmetic and personal care products, including chemistry and biochemistry, microbiology, molecular modeling, engineering, formulation science, technical packaging, and toxicology.

知识拓展

Unit Six
Beauty & Cosmetic Communicative English
美容化妆品交际英语

Learning Objectives　学习目标

In this unit you will be able to:
1. Learn how to greet the guests when they arrive at the beauty salon.
2. Learn how to do facial and body care services to customers.
3. Understand how to sell cosmetics and perfume to customers.
4. Understand how to put on make-up for customers.
5. Practice how to serve customers in a hair salon and a nail salon.

Lesson 23 Customer Reception
客户接待

Lead in 课前导入

Thinking and talking:
1. Do you know how to greet and serve customers when they arrive?
2. Do you know the process of making an appointment?
3. How about the process of after service?

Related words:

beauty salon	美容会所	receptionist	接待员
beautician	美容师	beauty advisor	美容顾问
manicurist	美甲师	massagist	按摩师
greetings	打招呼	front desk/reception desk	前台
make an appointment	预约	appointment book	预约簿
telephone reservation	电话预约	after service	售后服务

Task 1 Greetings & Understanding Needs

任务一 问候客户及了解需求

Key Sentences 焦点句型

1. Greetings 问候客人

Hello. /How do you do? /How are you? /Nice to meet you.
Welcome to our beauty salon. Can I help you?
If I'm not mistaken, you must be Miss Chen.
It's a great honor to meet you. /I have been looking forward to meeting you.
Do you have an appointment?

2. Showing the guests to the front desk/reception desk 带客人到前台休息

Please allow me to take you to the front desk, Mrs. Smith.
Please come with me/follow me to reception desk.

Lesson 23 Customer Reception

This way please.

3. Entertaining the guests 招呼客人

(1) 请客人坐下稍等片刻：

Please take a seat.

The beautician/manicurist/massagist is busy at the moment. I'll inform her/him right away.

(2) 询问客人是否喝水：

Would you like a glass of water? /Can I get you a cup of Chinese red tea?/How about a Coke?

A cup of coffee would be great. Thanks.

Alright, let me make some. I'll be right back.

离开一会：

Excuse me. I'll be right back.

Please excuse me for a moment.

4. Chatting with guests 与客人聊天

(1) Is this your first time to our beauty salon?

推荐服务项目：

(2) (Yes.) I recommend you an aromatherapy massage with lavender in our spa center. You'll feel good afterwards.

We also offer services such as facial care, body care, SPA treatment, make-up and manicure.

询问客人印象：

(3) (No. I've been here before.) What is the most satisfying service you have in our beauty salon?

(4) How do you think about our service?

聊生活、工作爱好：

(5) What do you like to do in your spare time?

What line of business are you in?

没听清，确认话意：

(6) Could you say that again/repeat that/speak a little more slowly, please?

5. Showing the guests around 带客人参观

(1) Since it's your first time here, let me show you around.

(2) Now we're in the reception room, where clients can make an appointment or check out. Our clients can take a rest and discuss their needs our with beauty advisor as well.

(3) This is our batching room for preparing materials.

(4) These are our beauty treatment rooms. We have two 3-bed rooms, two double-bed rooms, and we also have one single room for VIP clients.

(5) I hope you have a wonderful stay with us today.

Sample Dialogues 对话范例

Sample Conversation 1:

Situation: Rose is a receptionist of Angel Beauty Salon. She serves Mrs. Hu, who has an appointment with the beautician Shelly at 11:30.

(Mrs. Hu walks into the beauty salon.)

Rose: Good morning. Welcome to Angel Beauty Salon. Can I help you?

Mrs. Hu: Good morning. I have an appointment with the beautician Shelly at 11:30.

Rose: May I have your name, please?

Mrs. Hu: Hu. May Hu.

Rose: Let me check the appointment book... Ah yes, Mrs. Hu. Please take a seat. I'll tell Shelly you are here.

(Dials.)

Shelly: Hello.

Rose: Hello, Shelly, this is reception. Mrs. Hu is here. She has an appointment with you at 11:30.

Shelly: Oh, yes, that's right. Please bring her to Room 104.

Rose: All right. (Replace the phone.) Mrs. Hu, Shelly is expecting you in Room 104. Please come this way.

Mrs. Hu: Thank you.

Sample Conversation 2:

Situation: 一般的访客都有提前预约，但也有些"不速之客"。
The second visitor is Ms. Li. She doesn't have an appointment, but she wants to see Shelly.

(Ms. Li walks into the beauty salon.)

Ms. Li: Good morning.

Rose: Good morning. Oh, Ms. Li. How are you?

Ms. Li: I'm fine, thanks, and you?

Rose: Oh, busy as usual. Do you want to see Shelly today?

Ms. Li: Yes, please.

Rose: Do you have an appointment?

Ms Li: er... No, I don't.

Rose: Well, Shelly is busy at the moment. I'll check her schedule for you. Please sit down and have a rest.

Ms Li: Thank you.

(Dials.)

Shelly: Hello, Shelly's speaking.

Rose: Oh, Hello Shelly. It's reception again. I have Ms. Li here, Li Mei. She hasn't an appointment but she'd like to see you this morning. Is it convenient for you?

Shelly: Let me see... Well, hmmm, I'll be free at about 12:30. Can she wait?

Rose: (to Ms. Li) Shelly will be free about 12:30. Can you wait?

Ms Li: Oh that's fine. I'll wait.
Rose: (to Shelly) Shelly, Ms. Li will wait.
Shelly: Right. I'll inform you when I'm done then.
Rose: Thanks. (Replaces the phone.) (to Ms. Li) Ms Li, Please take a glass of water.
Ms Li: Thank you. You're so kind.
Rose: You're welcome. Would you like to read some magazines?

Key Words & Phrases   重点词汇

receptionist[rɪˈsepʃənɪst]   *n.* 接待员；传达员
beautician[bjuːˈtɪʃn]   *n.* 美容师
schedule[ˈskedʒuːl]   *n.* 计划（表）；时间表
convenient[kənˈviːniənt]   *adj.* 方便的

Practical Activities   实践活动

Situation:
It is afternoon. Ms. Weeks walks into Angel Beauty Salon. It's the first time for her to be here. Greet her and recommend some services to her.
Ms. Weeks finally chooses to have body care, but she has to wait for one hour.
Ms. Weeks is unwilling to wait. She feels bored and annoyed.
Try your best to entertain her.

Requirement:
Make sure to have your conversation with the spirit of love. The guest is picky. She needs to be served with extreme warmth and loving kindness.

Task 2   Make, Change or Cancel Appointment
任务二   制定、更改或取消预约

Sample Dialogues   对话范例

Dialogue 1: Make an appointment   制定预约
Situation：客人 Mrs. Reed 打电话到克丽缇娜美容院（Chlitina Beauty Salon）想预约美容师 Mary 这周做面部护理（have a facial），Mary 周一、二、四都约满了（fully booked），只有周三有空（available），而 Mrs. Reed 周三要上班，Mrs. Reed 想改约美容师 Jenny 周六，但 Jenny 周六休息（day-off），最后约定本周日下午找 Jenny 做面部护理。
Clerk: Good morning. This is Chlitina Beauty Salon. What can I do for you?
Mrs. Reed: Yes, this is Mrs. Reed. I'd like to have a facial with beautician Mary this week.
Clerk: Well, let's see. I'm afraid she is fully booked on Monday and Tuesday.
Mrs. Reed: How about Thursday?

Clerk: Sorry, but I have to say she is also occupied on Thursday. So, will Wednesday be OK for you, Mrs. Reed?

Mrs. Reed: I have to work on Wednesday. By the way, is beautician Jenny available on Saturday?

Clerk: I'm afraid she is day-off that day.

Mrs. Reed: Well, what about Sunday?

Clerk: Sunday. Let me have a check. Oh, great. Jenny will be available on Sunday afternoon this week.

Mrs. Reed: That's fine. Thank you, I'll come then.

Clerk: You're welcome.

Dialogue 2: Make and change an appointment 制定及更改预约

Situation: 客人 Ms. Wang 打电话到百莲凯养生馆（Balincan SPA Center）想约按摩师 John 今天下午做按摩，John 下午已约满，可约明天上午 10 点。Ms. Wang 询问后天下午 4 点 John 是否有空。职员查询预约簿过后，告知她 John 后天也忙，最后定下周一上午 10 点。客人又打电话，因下周一要出差（business trip），取消预约。双方重新约时间。

Ms. Wang: I wonder if I could come to see massager John this afternoon?

Clerk: Can you come here tomorrow morning at 10:00? I'm afraid John is occupied this afternoon.

Ms. Wang: Will he be free the day after tomorrow at 4:00 p.m.?

Clerk: Just a moment, please. I will check the appointment book.

(A few seconds later.)

Clerk: I'm sorry that he's also busy that day. How about 10:00 next Monday morning? The massager will be available.

Ms. Wang: Ok. Let's make it 10:00 next Monday morning.

客人又来电更改预约日期：

Ms. Wang: I'm sorry I'll have to cancel the appointment with massager John next Monday morning. I have a business trip that day.

Clerk: Oh, I'm sorry to hear that. Then let's reschedule the appointment.

Ms. Wang: All right.

Dialogue 3: Cancel an appointment 取消预约

Situation: 客人 Tracy 约了今晚 6 点半到 T-Nail Shop 做美甲，店员 Rebecca 在 5 点半提前给她打电话提醒。过了 7 点 Tracy 还没到，店员询问情况。Tracy 还在外面忙赶不过去，取消预约。

Rebecca: Hello. May I speak to Ms. Tracy Cooke, please?

Tracy: This is, who is calling please?

Rebecca: Hi, Tracy. This is Rebecca of T-Nail Shop. I'd like to remind you of your nail polish appointment tonight at 6:30.

Tracy: Oh, I nearly forgot. Thank you for reminding me. I'll be there on time. See you.

Rebecca: See you then.

(Just after 7 o'clock.)

Tracy: Hello.

Rebecca: Hello, Tracy. This is Rebecca. Are you on your way to our T-Nail Shop?

Tracy: No. Something has just stopped me. I am still outside busy with my work. I'm afraid I have to cancel the appointment tonight.

Rebecca: Oh, what a pity. Then let's reschedule the appointment next time.

Key Words & Phrases   重点词汇

appointment[əˈpɔɪntmənt] *n.* 预约
fully[ˈfʊli] *adv.* 完全地，充分地
book[bʊk] *n.* 书籍；卷；账簿 *vt.* 预定，预约
occupy[ˈɑːkjupaɪ] *vt.* 占据
available[əˈveɪləbl] *adj.* 可以见到的，有空的
check[tʃek] *vt.* 检查，查看
would like to 想要
make an appointment 预约
have to 不得不
by the way 顺便问一下
be occupied 没有空
be free/be available 有空
check the appointment book 查预约簿
make it 10:00 tomorrow morning 定在明天早上10点
reschedule the appointment 重新安排预约时间
cancel the appointment 取消预约

Practical Activities   实践活动

Translate the following sentences.

Points to Remember

1. Make eye contact and smile warmly as you greet and talk to the guest.
2. Your smile should be like "sweet scent" and bring "fragrance to the soul".
3. Think of how you can compliment the guest.
4. Look for something you like about the guest so that you can start to feel good about the guest. The guest will feel your warmth.
5. As you interact with the guest, say to yourself a thought, such as "I like this guest very much" or "I want to make this guest really happy". Focus on the thought during the interaction.
6. Look for ways to show love, care, warmth, and empathy during the interaction. The guest wants and needs to feel your love, care, and warmth, even though she/he might not admit it.
7. Your heart should be burning with loving kindness for the guest, and the guest should feel it. Let it show!

8. Stand straight but relaxed as you speak.

9. Look for information about the guest that can be used as guest history information in order to create a memorable experience.

10. Think about how you can make the guest happy as you interact with him/her.

11. Constantly look for ways to please the guest.

12. As you interact with the guest, listen to your heart about how you think the guest feels and what the guest wants. Then take action.

13. Try to practise the spirit of this quotation: "Shed the light of boundless love on every human being whom you meet."

Task 3 Post-service

任务三 售后服务

Sample Dialogues 对话范例

Dialogue 1: Fill in the feedback form and recommend a member card
Bella——a beauty receptionist   Jane——a regular client
Bella interviews Jane after her facial service.

Bella: Wow, Ms. Lin, your skin looks more radiant and moisturized than before. How do you feel? Are you satisfied with our service?

Jane: It's wonderful. I had a good sleep just now. The service and environment are so relaxing.

Bella: That's great. Please drink some scented tea.

Jane: Thank you. You're so kind.

Bella: It's my pleasure. Here is our post-service feedback form. Would you please fill in this form and sign your name here?

Jane: OK.

Bella: Will you consider to apply for a membership card? You can enjoy our member's 20% discount next time. This is the price list.

Jane: Oh, it's really worth buying. I'll buy one.

Bella: Thanks a lot. Will you pay by cash, credit card, Wechat or Alipay?

Jane: Wechat, please.

Bella: All right. We're looking forward to your coming next time. Bye.

Jane: Bye.

Dialogue 2: Post-service Interview
Bella——a beauty receptionist   Jane——a regular client
Bella interviews Jane on the phone.

Bella: Hello. This is Angel Beauty Salon. Is that Ms. Lin?

Jane: Yes, it's me.

Lesson 23 Customer Reception

Bella: May I have a post-service interview with you? It won't take you too much time.
Jane: OK, no problem.
Bella: How do you feel about our beautician's service last week?
Jane: The beautician was very conscientious and I did a good relaxation that day. Thank you very much.
Bella: You're welcome. How about our products? Do you remember to use our masks regularly at home?
Jane: Yes, I do it twice a week. I feel my skin is more moisturized than before.
Bella: Great. Will you come at the same time this week?
Jane: Yes, if I'm available then.
Bella: Ok. I'll remind you on Wechat this Friday. See you then. Bye.
Jane: Thank you. Bye.

Key Words & Phrases 重点词汇

radiant['reɪdɪənt] *adj.* 辐射的；容光焕发的；光芒四射的
moisturize['mɔɪstʃəraɪz] *v.* 增加水分；变潮湿
scented tea 花茶，花果茶
post-service 售后服务
feedback['fiːdbæk] *n.* 反馈；成果，资料；回复
membership['membərʃɪp] *n.* 资格；成员资格；会员身份
conscientious[ˌkɑːnʃi'enʃəs] *adj.* 认真的；尽责的；本着良心的；小心谨慎的
relaxation[ˌriːlæk'seɪʃn] *n.* 放松；缓和；消遣

Practical Activities 实践活动

Part A Matching the words with the pictures.

Feedback Form   Wechat Payment   Membership Card   Cash Payment   Credit card Payment

Part B Listening practice.
Dialogue 1:
M: Hello, Bill Berton speaking. What can I do for you?
F: Hello, Mr. Berton, this is Jenny Jenkins of Bradford and Sons returning your call. I'm sorry 1. _____ when you called my office this morning. My secretary said you called concerning our meeting next Tuesday?

M: Yes Ms. Jenkins. Thank you for returning my call. I'm glad to finally 2. _____ . I want to let you know I will not be able to 3. _____ next Tuesday. I would be out of town that day. Is there any possibility we can move the meeting to Monday?

F: I am sorry. I'm afraid I am completely booked on Monday. Would it be possible to postpone until you return?

M: Oh, dear. I was 4. _____ taking care of our meeting before I leave. But I suppose I can 5. _____ a few things. Yes, we can arrange something. I will be back Thursday morning. What about Thursday afternoon? Would that work for you?

F: That should be fine, shall we say about two o'clock?

M: Perfect! I look forward to seeing you at two o'clock next Thursday afternoon. If you need to change the time, please feel free to call me 6. _____ .

F: Thanks Mr. Burdon. I'll see you on Thursday.

Dialogue 2:

F: Hello.

M: Hello, is Doras available?

F: This is Doras, who is calling please?

M: Hi, Doras. This is Mike calling from Parker's Beauty. I'm calling to 1. _____ for tomorrow morning at 9 a.m. with Doctor Parker.

F: Oh, I almost forgot. Thank you for calling to remind me. Actually, I do need to change the time of my appointment. I have a 2. _____ , and I can't make it that early.

M: If I 3. _____ that later slot, would that work out?

F: It would have to be after lunch. Do you have anything available about 2 o'clock?

M: Sorry. The only opening we have after lunch is 1:15, but I might be able to 4. _____ at 4:00. Would that be a better time?

F: That's all right. I think I should be able to make it at 1:15. Can you 5. _____ that time slot?

M: No problem, I have your appointment changed from tomorrow morning to tomorrow afternoon at 1:15.

F: Wonderful. Thanks very much.

Lesson 24 Skin Care Service
皮肤护理服务

Lead in 课前导入

Thinking and talking:
1. Are you clear about your skin type?
2. Do you know how to communicate with customers when doing facials for them?
3. Do you know the skills of introducing facial services and products to customers?
4. How many types of massage do you know? How about the skills?

Related words:

Skin care products	护肤品	eye gel	眼部胶
skin care	护肤	facial mask/masque	面膜
cleansing milk/facial cleanser/face wash	洗面奶	eye mask	眼膜
toner/astringent	爽肤水	lip care/lip balm	护唇用
firming lotion	紧肤水	lip coat	口红护膜
toner/smoothing toner	柔肤水	facial scrub	磨砂膏
moisturizers and creams	护肤霜	Pore	毛孔
moisturizer	保湿	stripper	剥去
sun screen/sun block	隔离霜,防晒	(deep) pore cleanser/ stripper pore refining	去黑头
whitening	美白		
lotion	露	exfoliating	剥落
cream	霜	exfoliating scrub	去死皮
day cream	日霜	body lotion	润肤露(香体乳)
night cream	晚霜	hand lotion/moisturizer/hand cream	护手霜

Functions 护肤品功能

multi-	多元	facial	脸部用
normal	中性皮肤	hydra-/hydration	保湿用
oily	油性皮肤	scrub	磨砂式（去角质）
dry	干性皮肤	waterproof	防水
combination/mixed	混合性皮肤	fast/quick dry	快干
balancing	平衡	after sun	日晒后用品
oil-control	抑制油脂	alcohol-free	无酒精
nutritious	滋养	essence	精华液
anti-	抗、防	essential oil	香精油
anti-wrinkle	防皱	toner/toning lotion	化妆水
anti-aging	抗老	gentle	温和的
firm	紧肤	foam	泡沫
sensitive	敏感性皮肤	long lasting	持久性
milk	乳	repair	修护
pack	剥撕式面膜	treatment	修护
peeling	敷面剥落式面膜	active	赋活用
acne/spot	青春痘用品	day	日用
clean-/purify-	清洁用	night	夜用
whiten	美白用	revitalite	活化
sunblock	防晒用	nutritious	滋养

Task 1　Understanding Your Skin

任务一　认识你的皮肤

Sample Dialogues　对话范例

Dialogue：Mary is a beauty adviser. Customers Linda, Dianna, Lily, Daisy are consulting skin problems with Mary.

Mary：Hello. I'm Mary, your beauty adviser. As we know, people have different types of skin. Do you know what types of skin are there?

Linda：There are five types of skin：oily, normal, mixed, dry and sensitive.

Mary：Absolutely right. We can check your skin type with this skin analysis apparatus. Linda, do you know your skin type?

Linda：I have oily skin. Just like many youngsters, I have acne on my forehead and nose.

Mary：You're prone to form acne skin and you'd better pay more attention to skin cleansing and choose light lotion or cream. Remember to eat less spicy food. Dianna, how about you?

Dianna：I'm prone to have red spots when I try some new products or eat seafood, such as little shrimps.

Mary：Obviously, your skin is sensitive You need to be cautious when using new skin care products and eat less seafood.

Lesson 24 Skin Care Service

Lily: Maybe my skin is dry. I put on much lotions every day, but my skin is still very dry. What should I do?

Mary: You should add some creams and moisturizers to keep the balance of water and oil on your skin.

Daisy: I think I have mixed skin. My T-zone is always oily while cheeks are dry. What should I do?

Mary: It's very important to replenish water and keep moisture at any time, especially for people with dry and mixed skin. It's better to have moisturizing masks daily or at least three times a week.

Daisy: Thank you so much.

Mary: My pleasure.

Key Words & Phrases   重点词汇

spicy['spaisi] *adj.* 辣的
shrimp[ʃrimp] *n.* 虾
analysis[ə'naelasis] *n.* 分析，解析
apparatus[ˌæpə'reitəs] *n.* 器官，装置，机构，组织，仪器
skin analysis apparatus   皮肤分析仪
replenish[rɪ'plenɪʃ] *vt.* 补充，再装满；把……装满；给……添加燃料
moisturize['mɔɪstʃəraɪz] *vt.* 给……增加水分，使……润
T-zone   T字部位
cheek[tʃi:k] *n.* 面颊

Practical Activities   实践活动

Part A Look at the pictures and write the skin types.

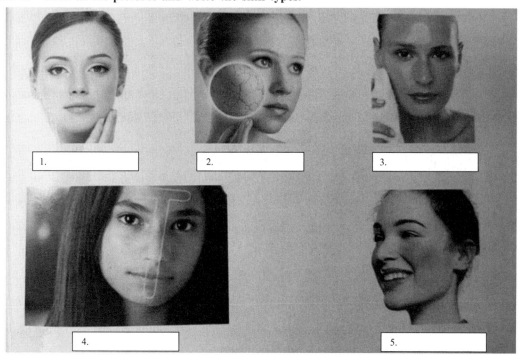

1.
2.
3.
4.
5.

Part B Work in pairs and fill in the blanks about skin types.

Skin Types	Symptoms	Solutions
Oily skin		
Sensitive skin		
Dry skin		
Combination skin		

知识拓展

Skin

Task 2 Facial Care

任务二 脸部护理

Sample Dialogues 对话范例

Dialogue 1:

Amy——a beauty adviser   Betty——a customer

A: Good evening, Madam. May I help you?

B: I want a facial. But this is the first time I've come here, so can you tell me how you do it?

A: Sure. Most facials start with thorough cleaning. Then we usually use a toner to invigorate the skin, followed by an exfoliation treatment—a peeling mask or scrub that removes the dead cells that make the skin look dull. After that, we'll massage your face and neck with oil or cream to improve the circulation and relieve the tension, followed by a mask to moisturize and soften the skin.

B: That's exactly what I want. I'll treat myself to a facial message today.

A: You deserve it. And we're having a special on facials this weekend.

B: My good luck. the freezing weather recently has ruined my complexion, you know, so it's time I had a facial.

A: Not to worry. I'll also take care of your eyebrows and eyelashes. I promise you'll feel great after the treatment.

Dialogue 2:

Amy——a beautician   Betty——a customer

A: Let's start with the cleaning masque.

B: What are the ingredients in your cleaning masque?

A：It's a combination of oatmeal, yogurt and honey.

B：What are they good for?

A. Oatmeal is a very gentle exfoliator that can remove dead skin from your face.

B：Uh-huh. What about yogurt and honey?

A：Yogurt can soften the skin, and honey is one of the best natural humectants.

B：I see. Will you do the "rose petal facial steam" for me?

A：No. I plan to do the peppermint and rosemary facial steam because the combination of these two plants will soothe tired muscles.

B：Great.

A：Now, no more talking and relax for a while. I'll massage your scalp.

B：OK. This is the part of facials that I like best. I'll enjoy it!

Key Words & Phrases   重点词汇

invigorate[in'vigəreit]   *vt*. 增添活力
improve the circulation   促进血液循环
relieve the tension   消除紧张
special   特价
eyebrows   眉毛
eyelashes   睫毛
restorative treatment   提神补元疗程
restorative[ris'tɔrətiv]   *adj*. 恢复健康和体力的，恢复的
scalp[skælp]   *n*. 头皮
oatmeal['əut,mi:l]   *n*. 燕麦
humectant[hju:'mektənt]   *n*. 湿润剂
peppermint['pepə,mint]   *n*. 薄荷
rosemary['rəuzmeri]   *n*. 迷迭香

Practical Activities   实践活动

Part A   Translate the following key sentences.

1. Do you want a face massage?
2. I need to have a facial to relax a little bit.
3. Do you offer facials?
4. My skin looks so dull, what kind of facial treatments will improve it?
5. My face has started breaking out. Can I still have a facial?
6. My lips are peeling. Do you have a special treatment for the lips?
7. Must I clean my face before doing the steam?
8. Is "lavender facial steam" good for acne-prone skin?
9. You should use this facial masque often.
10. I'm sure the facial massage will relax your tired skin and flesh.
11. Would you try a special mask to get rid of the filth and grease?

12. Now, shall we use the youth cream?
13. It's time to smooth your beautiful face with sun protection cream.

Task 3　Introducing Service & Products

任务三　服务及产品推荐

Key Sentences　焦点句型

1. Do you carry alcohol-free toners?
你们有没有不含酒精成分的化妆水？
2. I've started getting wrinkles. Do you have any anti-wrinkle products?
我开始有皱纹，你们有卖防皱的产品吗？
3. My skin is very sensitive. I can't use moisturizer that contains fragrance.
我的肌肤很敏感，没办法用含香味的保湿霜。
4. Do you carry anything that can improve dry and cracked lips?
你们有没有可以修复干裂嘴唇的产品？
5. The eye cream I bought last time was too greasy, do you have a lighter one?
我上次买的眼霜太油腻了，有比较清淡的。
6. I heard that your anti-aging gel works really well. I'm interested in buying one bottle.
我听说你们的抗老化精华露非常棒，我想买一瓶。
7. Do you sell moisturizer that is specially made for the neck?
你们有颈部专用的保湿霜吗？
8. I have oily skin, so I want something that is light but can still keep enough moisture for the winter.
我是油性皮肤，所以我需要那种比较清爽但在冬季又足以保湿的产品。
9. My skin looks dull. What would you recommend that I use?
我的皮肤看起来很淡，你会建议我用些什么产品？
10. I feel my pores enlarging. Is there anything I can use to shrink them?
我觉得我的毛细孔变大了，有没有什么产品可以修复？
11. Do you have products to ease breakouts?
你们有治痘痘的产品吗？
12. You can try to use this honey and almond enriched facial scrub. It will exfoliate the dead skin.
你可以试用这种富含蜂蜜及杏仁成分的磨砂洁面霜，它可以去除老化的皮肤。
13. The skin under eyes is very sensitive. You shouldn't apply regular moisturizer there.
眼睛下围的肌肤很敏感，你不应该在那儿擦一般的保湿霜。
14. If your skin is sensitive, you'd better do a "patch-test" before applying the products on your face.
如果你的皮肤很敏感，你最好在脸部使用产品前先做一下"过敏反应测试"。

15. This is a sample of our "anti-wrinkle eye cream". Would you like to take it home and try it?
这是我们的"防皱眼霜"试用品，你想带回家试用吗？
16. I think you might need this moisture cream in winter.
我想你冬天可能会需要用到这种保湿霜。
17. I'll try it first. If I like it, I'll come back.
我先试用一下，如果喜欢的话，我会再回来买。
18. What can you do about my wrinkles?
你有什么办法消除我的皱纹呢？
19. Is there some face powder to hide my freckles?
有没有蜜粉可以掩盖我的雀斑？
20. Could you recommend a kind of cleansing milk to me?
你能给我推荐一种洗面奶吗？
21. I want a tube of granular cleanser.
我要一管磨砂膏。
22. The mask suits for dry skin.
这面膜适合干性皮肤。
23. My facial skin has been very dry these days. Do you have any moisturizing lotion?
最近我脸上的皮肤很干燥，你们有保湿霜吗？
24. There isn't the date of production on this box of nutrition cream.
这盒营养霜上面没有生产日期。

Sample Dialogues 对话范例

Dialogue 1: Introducing service

Amy——a beauty adviser   Bella Liu—a customer

A: Miss Liu, you look exhausted. Would you like to try our "restorative treatment"?
B: What kind of treatment is that?
A: It's a specially designed treatment to release tense muscles, and it includes a hydrating facial and a hair and scalp massage.
B: How long does it take?
A: About 100 minutes.
B: How much is it for that?
A: $200. But, trust me, it's worth trying. After you finish the session, you'll feel terrific!

Dialogue 2: Knowing skin types & introducing products

Anna——a customer   Betty—a counter salesperson

Anna: Hi, I would like to know more about your moisturizers, could you tell me more about them?
Betty: Sure. Do you know your skin type?
Anna: I'm not sure My T-zone gets oily easily, but my cheeks are dry in the winter.

Betty: I see. This "Creme do Olives" is our bestseller for combination skin.
Anna: What is good about it?
Betty: It contains a very powerful antioxidant lotion made of olive-leaf extracts and Mediterranean herbs.
Anna: Won't it be too oily for my T-zone?
Betty: Not at all. It's rich, but not heavy.
Anna: Could I try some on my hand?
Betty: Of course. See. it can be absorbed quickly.
Anna: Hmm, I don't feel greasy at all, and my skin seems to instantly become smoother.
Betty: Amazing, right? Do you need anything else?
Anna: Yeah. I always get rough skin in winter. Why?
Betty: Well, since the weather becomes cool and dry, your skin gets dehydrated easily.
Anna: So how can I keep it soft and supple?
Betty: You can use a gentler, creambased facial cleanser, and always remember to follow that with a thick moisturizer.
Anna: Do you sell them?
Betty: Yes, we do. This is our "ultra-moisturizing cleansing cream".
Anna: Ha, just suits my needs.

Key Words & Phrases　重点词汇

moisturize[ˈmɔɪstʃəraɪz]　vt. 给……增加水分,使……润
T-zone　T字部位
antioxidantlotion　抗氧化剂
olive-leaf　橄榄叶
Mediterranean[ˌmedɪtəˈreɪnjən]　地中海
dehydrated　脱水的
soft and supple[ˈsʌpl]　柔嫩细致
ultra[ˈʌltrə]　超效

Practical Activities　实践活动

Part A　Work in pairs and match the sentences with the following dialogues about knowing skin types & introducing products.

1. I'm very sensitive to fragrance.

2. What would you recommend?

3. What kind of skin do you have, normal, oily or dry?

Lesson 24　Skin Care Service

Dialogue 1：

A：I'd like to buy a bottle of cleansing milk. Would you kindly give me some suggestion?

B：I'd love to. _____ .

A：My skin is always dry.

B：Then I recommend the American brand "Elisabeth Arden". It contains honey, skin milk and sesame oil. It will improve your skin.

A：Really?

A：我想买瓶洗面奶，您能给我点建议吗？

B：很愿意。您属于哪种皮肤？中性的，油性的，还是干性的？

A：我的皮肤总是干干的。

B：那我推荐美国的"伊丽莎白雅顿"牌。它含有蜂蜜、肌肤乳液和芝麻油，会改善您的皮肤。

A：真的吗？

Dialogue 2：

A：I want to buy some cleansing milk. _____ .

B：Your complexion is on the oily side. I suggest you using cleansing gel.

A：Anything that'll keep my skin clean.

B：How about this one? It cleans thoroughly without striping your natural protective oil. The gentle formula keeps skin soft and healthy.

A：Hm. the smell is too strong, _____ .

B：We've also got a fragrance-free one, especially for sensitive skin. I'm sure you'll like it. This line of products is fragrance-free. We have facial masks, moisturizing lotions, eye creams and tonics.

A：OK, I'll try the moisturizing lotion and cleansing gel first.

B：Thank you very much, sir. Here are some samples of our products. Do try them out.

A：Thank you so much.

A：我想买洗面奶，你可以推荐一款给我吗？

B：您的皮肤是油性的，我建议您用洁面膏。

A：任何能保持我面部清洁的都成。

B：这个怎样？它能彻底清洁您的皮肤，又不会洗去保护表皮的天然油分，配方温和，能保持皮肤柔软及健康。

A：唔……这个味道太浓，我对香味是非常敏感的。

B：我们有一款无香味的，是专为敏感皮肤而设，相信您一定会喜欢。这个系列的产品都是无香味的，包括面膜、补水乳霜、眼霜及爽肤水。

A：好的，我先试试补水乳霜及洁面膏吧。

B：谢谢你，女士。这儿有一些我们其他产品的试用装，请拿回去试试。

A：十分感谢。

Part B Translate product labels.

1. CLINIQUE
all about eyes rich

2. CETAPHIL
Gentle Skin Cleanser

3. La Mer
The Concentrate

4. Fresh
Black Tea Instant
Perfecting Mask

5. ESTEE LAUDER
Advanced Night Repair Eye Recovery Complex

Task 4 Massage & Body Care

任务四 按摩及身体护理

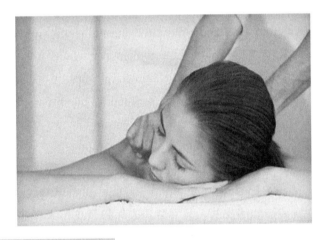

Sample Dialogues 对话范例

Dialogue 1: Neck & shoulders' massage

Mary—a massagist   Shelly—a customer

Mary: Good afternoon, miss. What can I do for you?

Shelly: I'd like a neck and shoulders' massage.

Mary: You look exhausted. What's the matter with you?

Shelly: I've been stayed up late recently. I'm so tired and my body feels very week.
Mary: I see. I'll give you a Swedish body massage. Now, please lie down. Let's start with a neck massage. If you feel it's too strong, just let me know.
Shelly: Oh, it works. Could you please pinch my shoulders harder?
Mary: No problem. Are you comfortable with my force?
Shelly: Yes, I feel much better now. Thank you.
Mary: You're welcome. Just relax.
Shelly: What kind of oil is that? It smells so good.
Mary: It's mix of lavender, rosemary and grape-seed essential oil. It can relax your tense muscles.
Shelly: That's great. Can you give me some suggestions?
Mary: Sure. Don't stay up too late and do regular exercise. Try to stay away from computer and mobile phone as much as possible.
Shelly: Thank you very much.

Dialogue 2: Back massage

Nancy has reserved a body massage. She goes to a massage parlor.

A——Massagist Lily   B——Nancy

A: Hello Nancy. Your appointment today is a back detox massage.
B: Yes. This is my first time doing this massage. Would you please explain it?
A: The purpose of this massage is to increase smooth flow of the lymphatic system and eliminate excess water and toxins in the body. It takes about one hour. Let's begin.
B: Oh, it feels comfortable when you are doing a circular massage.
A: It is done. Now I am pressing your skin to help you fully relax.
B: OK. Thanks.
A: The next step is a whole back massage for about 30 minutes. I am now applying a lymphatic detox technique to massage you, removing toxins from your body.
B: It aches, but in a comfortable way.
A: Now I am massaging your waist at the back. The force is slightly stronger when l am pressing on your spine. Now, how do you feel? Would you like strong pressure?
B: I think it's alright.
A: I'm now massaging an important acupoint in your waist. Is it comfortable?
B: It's a little painful. Softer, please.
A: Alright.
(About half an hour later.)
A: The massage is over. How do you feel now?
B: I feel much more relaxed.
A: That's good. Please put on your clothes. Keep warm.
B: Thank you.

A: You are welcome.

Dialogue 3: Body shaping program

Ms. Chen is obese after childbirth and hopes to improve her figure in a beauty salon.

A——Beauty consultant   B——Ms. Chen

A: Hello, Ms. Chen. Can I help you?

B. Yes. Look at my belly. It is big after I gave birth to my child.

A: We have a belly shaping program. It helps to shape your belly.

B: Is the process comfortable?

A: Not really. We need use the weight reducing apparatus to eliminate excess fat during the process. It is a bit tiring.

B: Is there any program which is more relaxing?

A: Yes, we have a meridian slimming program. It removes the excess fat by meridians conditioning. It is more comfortable.

B: What's the price?

A: 3,999 RMB.

B: How long does the whole process take?

A: About two months.

B: Sounds good. I think meridian slimming fits me more. Thank you.

A: You are welcome.

Key Words & Phrases 重点词汇

exhausted[ɪgˈzɔːstɪd] *adj.* 筋疲力尽的，疲惫不堪的；耗尽的，枯竭的

stay up   不睡觉，熬夜

pinch[pɪntʃ]   *v.* 捏；夹紧；夹（脚）

detox[ˈdiːtɑːks]   *n.* 排毒

lymphatic system   淋巴系统

eliminate[ɪˈlɪmɪneɪt]   *vt.* 消除；排除

toxin[ˈtɒksɪn]   *n.* 毒素；毒质

circular massage   打圈按摩

ache[eɪk]   *v.* (持续的)疼痛；渴望；哀痛，怜悯；感到痛苦

waist[weɪst]   *n.* 腰，腰部

spine[spaɪn]   *n.* 脊柱，脊椎

acupoint[ˈækjʊpɒɪnt]   *n.* 穴道，穴位

belly[ˈbeli]   *n.* 腹部

apparatus[ˌæpəˈrætəs]   *n.* 装置，设备；仪器；器官

meridian slimming program   经络减肥计划

meridian[məˈrɪdiən]   *n.* 子午线，经线；中医经脉

Lesson 24 Skin Care Service

Practical Activities 实践活动

Part A Write down the English or Chinese of body parts.

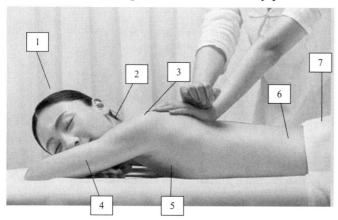

1. 头部 _____
2. 颈部 _____
3. 肩部 _____
4. 手臂 _____
5. _____ breast
6. _____ waist
7. _____ hip

Part B Translate the following table of beauty programs.

No.	Program	Effect
1	Spot lightening	Repair spotted skin
2	Head treatment	Refresh your mind; improve sleep quality
3	Back treatment	Dredge internal channels; relieve stress
4	Belly shaping	Make the belly smaller
5	Breast care	Maintain the shape of breasts
6	Leg shaping	Reduce fat; shape legs
7	Hip shaping	Lift the hips
8	Gastrointestinal conditioning	Relieve stomach discomfort; enhance intestinal function
9	Stretch mark repairing	Repair stretch marks after childbirth
10	Meridian slimming	Remove excess body fat by meridians conditioning

Part C Translate the following massage types.

1. Thai Massage _____
2. Swedish Massage _____
3. Hot Stone Therapy _____
4. Body Scrub _____

5. Bath in milk _____
6. Mineral mud bath _____
7. TCM Spa (Traditional Chinese Medicine) _____
8. Aromatherapy _____

知识拓展

Lesson 25 Selling Cosmetics & Perfume
销售化妆品及香水

Lead in 课前导入

Thinking and talking:
1. Do you know the skills of selling color cosmetics to customers?
2. How about the process of selling perfume?

Related words:

cosmetic counter	化妆品专柜
department stores	百货商店
salesperson	销售人员
shopping guide	导购员
try products out for customers	为顾客试用产品

重点词汇

Lesson 25　Selling Cosmetics & Perfume

Task 1　Selling Cosmetics

任务一　销售化妆品

Key Sentences　焦点句型

1. I want to buy a new lipstick. Do you know any good brands.
我想买一支新的唇膏，你知道哪个牌子比较好用吗？
2. How much for this black mascara?
这支黑色的睫毛膏要多少钱？
3. What is included in your gift package?
你们的赠品包括了什么？
4. With this coupon, you'll give me a 15％ discount for purchasing night cream, right?
有这张折价券，我买晚霜会是八五折的优惠，对吗？
5. Excuse me, how can I get the extra free handbag?
对不起，我要怎样才可以得到那个额外的免费手提袋？
6. I'm sorry. The gift package is exclusive for customers who purchase our new foundation.
很抱歉，这项礼品只赠送给购买我们新上市粉底霜的客人。
7. There are two different sets of eye shadows in our gift package, which one would you like to have?
我们的赠品中有两种不同组合的眼影组，您想选择哪一组？
8. I'm not sure if this color matches my skin tone, can I try it first?
我不确定这个颜色是否适合我的肤色，我可以试用看看吗？
9. My skin is very sensitive. Do you have trial samples that I can try first?
我的皮肤很敏感，有试用品可以让我先试用一下吗？
10. If my skin is allergic to the product, can I bring it back for a refund?
如果我的皮肤对那个产品过敏，我可以拿回来退吗？
11. I'm sorry, if you have any allergic reactions, we can only promise exchange.
很抱歉，如果您有任何过敏的反应，我们只能接受换货的服务。
12. Excuse me, could you show me how to put on this eye shadow?
对不起，你可以示范给我看怎么用这些眼影吗？
13. If I don't like the lipstick's color that is in your gift package, is it possible to replace it with a different color?
如果我不喜欢赠品中的口红的颜色，我可以换一个不同颜色的吗？
14. I'm sorry. You can not change any item from the gift package.
对不起，赠品中的产品是不可以替换的。
15. What color do you usually wear for the eye shadow?
您通常使用什么颜色的眼影？
16. How about wearing fake eyelashes instead of using an eyelash curler?

不用睫毛夹，试试假睫毛如何？

17. Would you prefer a lipstick in lighter shade?

用淡一点颜色的唇膏好吗？

18. It is a newly imported product with the transparent touch.

这是最新进口的产品，具有透明感。

Sample Dialogues 对话范例

Dialogue 1：At the Perfect Diary cosmetics counter 在完美日记专柜

Sales：May I help you，Miss?

Customer：Yes. I'd like to look at lipstick and eyeshadow.

Sales：We have a beautiful selection of eye shadows this fall. Look at the colors. Aren't they beautiful?

Customer：But they're brown. I prefer a purple set.

Sales：Why don't you wear reddish brown eyeshadow for a change? We also have a lipstick to go with it.

Customer：Can I try it?

Sales：Of course. Have a seat，please. Now，here is the mirror. How do you like it?

Customer：Not bad. Actually，it makes me look younger. I like it.

Dialogue 2：In Watsons cosmetics shop 在屈臣氏化妆品专卖店

Addison (A)：How can I help you?

James (J)：I'd like to buy some perfume for my girlfriend.

A：Do you know what kind of scent she usually wears?

J：She usually doesn't wear anything but a few drops of Chanel No. 5. But I'd like to buy her a new fragrance.

A：Ok，here are some of our most popular perfumes.

J：Which one would you recommend?

A：Personally，I quite like the new perfume by Clinique. It's a subtle flowery scent. What do you think?

J：That smells great. I'll take one bottle，please.

A：Would you like to buy any other cosmetics for your girlfriend? We have a full range of products from cosmetics to skin cleansers and moisturizers.

J：That's OK. She normally just wears a little foundation and some loose powder，and I wouldn't know what shade to buy.

A：How about some lipstick? Every woman needs a nice tube of red lipstick.

J：She doesn't usually wear lipstick. She thinks it makes her nose look too big.

A：How about some mascara? That will make her eyes look bigger.

J：No，thank you. She has big enough eyes as it is.

A：I think that she would like some whitening cream.

J：No thanks. Western women usually try to make their skin darker，not lighter.

A：Will that be all then?

Lesson 25　Selling Cosmetics & Perfume

J: That will be all. You've been very helpful, thanks.

Key Words & Phrases　重点词汇

selection[sɪ'lekʃn]　*n.* 选择，挑选；选集；精选品
fall[fɔːl]　*v.* 落下；跌倒；（雨或雪）降落　*n.* 落下；跌倒；秋季
subtle['sʌtl]　*adj.* 微妙的；精细的；敏感的
scent[sent]　*n.* 气味；嗅觉
recommend[ˌrekə'mend]　*v.* 推荐，介绍

Practical Activities　实践活动

Work in pairs and match the sentences with the following dialogues.

1. Do you have any particular brand in mind?
2. Have you heard about our special promotion this month?
3. that's the very thing I need.

Dialogue 1

A: I'm looking for some lipsticks. Do you still have some in peach rose?

B: Oh, yes. That is a beautiful color. It has been a very popular lipstick this season. I have just two left.

A: Great. I'll take one.

B: _____ If you purchase at least 200 yuan in any L'Oreal products, you will receive this black tote with a sample of blush, mascara and two shades of eye shadow.

A: Wow. That sounds like a bargain. I'm running low on facial moisturizer and powder. Could you ring those up for me too along with the lipstick?

B: I'd be glad to. Do you need anything else?

A: That's all.

A: 我想买口红。有没有桃红色的？

B: 有。桃红色挺漂亮的，是这个季节的流行色。还剩两支。

A: 太棒了，我要买一支。

B: 您有没有听说过，我们这个月的特别促销？您要是买欧莱雅产品超过200元，就可以获赠这个黑色袋子，里边有一盒样品腮红、睫毛膏以及两种色调的眼影。

A: 哇！听起来可真划算。我的面霜和粉用得都差不多了。可不可以把这两样跟口红一块儿算。

B: 可以。其他的还要吗？

A: 不要了。

Dialogue 2

A: Welcome, Miss. May I help you?

B: I hope so. I want a lipstick.

A: _____

B: I like Lifei very much.

A: We have different shades of Lifei lipstick. May I know what color you usually wear?

B: Pink. But today, I'm thinking of buying one in a dark shade. You know I will be a teacher next month. I wish to look more serious.

A: Yes, I see. How do you like this one?

B: Not too bad. May I have a try?

A: Certainly.

B: Mmm... It's still too bright. Any darker shades?

A: Not from the Lifei group, I'm afraid.

B: Well, any brand will do so long as I can get the right color.

A: How about this one, then? It's with much transparent touch.

B: Oh, _____

A: Anything else?

B: Nothing more, thanks. How much is it altogether?

A: Let's see what that'll all cost 25 yuan for this lipstick, 36 yuan for the face powder, 18 yuan for the perfume. So they add up to 79 yuan.

B: Here you are.

A: (Holding the note) Well, a one-hundred-yuan note. 79 yuan from 100 yuan leaves 21 yuan. See if it's all right.

B: Exactly.

A: Here is your receipt and change.

B: Thank you very much.

A: My pleasure.

A: 欢迎光临，小姐。要买东西吗？

B: 是的，我想买口红。

A: 你喜欢什么牌子的？

B: 我很喜欢丽妃。

A: 我们有各种不同颜色的丽妃口红，请问你平常都用什么颜色的？

B: 粉红色的。但是，今天我想买深色的。你知道，下个月我就要当老师了，我希望看上去严肃些。

A: 我明白了。你觉得这种怎么样？

B: 还不错，我可以试试吗？

A: 当然可以。

B: 嗯，还是太浅了，再深一点的有吗？

A: 恐怕丽妃系列中没有。

B: 哪一种都可以，只要颜色合适就行。

A: 那么这种怎么样？它富有透明感。

B: 噢，正是我想要的。

A: 还要买点别的吗？

B: 就买这些，谢谢。总共多少钱？

A: 让我看看：口红 25 元、粉扑 36 元，香水 18 元，所以总共是 79 元。

B：给你钱。

A：(拿着钱) 哦，100元的票子，100元减去79元剩余21元，您看看对不对。

B：一点不错。

A：这是给您的收据和找零。

B：非常感谢。

A：别客气。

Task 2　Selling Perfume

任务二　销售香水

Key Sentences　焦点句型

1. The perfume is particularly elegant in smell.
这种香水香味特别高雅。

2. That's a nice perfume.
那是不错的香水。

3. Where's my cologne?
我的古龙水呢？

4. Can I recommend this Coco Chanel perfume?
我能为您推荐可可香奈儿香水吗？

5. It's advertised everywhere, very popular.
它的广告随处可见，很受欢迎。

6. It's the thing that would take a lady's fancy.
这正是女士中意之物。

7. I want to buy some perfume for my girlfriend.
我想买香水给我女朋友。

8. Thank you for your advice.
谢谢你的建议。

9. I'd like to buy some perfume.
我想买一瓶香水。

10. I like Revlon perfume better.
我比较喜欢露华浓香水。

11. Do you need perfume?
需要香水吗？

12. I'm much obliged to you for your patient explanations and introduction.
非常感谢你耐心的解释和介绍。

13. My boyfriend bought me a bottle of perfume.
我男朋友给我买了一瓶香水。

Sample Dialogues 对话范例

Dialogue 1

A：Hi, I'm looking for some perfume for my fiancee.

B. What do you have in mind?

A：I'd like something that's not too strong. She likes softer smelling perfume.

B：OK, here are three different bottles. Which one of these do you like best?

A：Let me see... Oh, I like that! What's the name of it?

B：That's "Poison", a famous brand.

A：Its name will frighten everybody.

B：You can say that again. But it's quite popular.

A：OK. I'll take it.

Dialogue 2

A：Welcome to our counter. What can I do for you, Miss?

B：I'd like to buy a bottle of perfume, the best one, please.

A：OK. This is made in France. It has a European jasmine fragrance.

B：Mmm, it's a well-known make. I'll have one bottle.

A：Anything else?

B：Do you have any skin tonic cream?

A：Yes. This is a pearl cream made in Shanghai. It conserves skin efficiently.

B：Does it really?

A：Sure

Dialogue 3

A：This is a bottle of Parisian perfume. It is very fragrant and will keep indefinitely, and has the reputation for quality that is unequalled.

B：Yes. I do agree.

A：Do you want a bigger bottle?

B：No, this will do for now. I'd like to test this one first, if you don't mind.

A：Not at all. Try this one, sir. The scent is really soft and alluring.

B：Hmm. It smells wonderful. But I am looking for something I can wear every day. Can you recommend anything?

A：Then we have the perfume you want. It's also from France. Look here it is.

B：This fragrance is also good. OK. I just take this bottle.

A：Here you are.

Dialogue 4

A：What kind of perfume do you have?

B：We have both Chinese and western perfume. Which do you like better?

A：Chinese, please. I want to get one bottle for my girlfriend as a souvenir gift. Please show me something of good quality.

B：How about this Qingfei perfume? It has won a national super-quality certificate. It's advertised on TV, very popular.

A: Really? What makes it so special?

B: The elegant smell. It's just the thing that would take a lady's fancy.

A: Then I will get a bottle. How much is it?

B: 100 yuan. So the total is 130 yuan.

A: Here you are.

Key Words & Phrases 重点词汇

fiancee[ˌfiːɑːnˈseɪ]   n. 未婚妻
jasmine[ˈdʒæzmɪn]   n. 茉莉
tonic[ˈtɑːnɪk]   n. 补药；主调音或基音   adj. 滋补的；声调的；使精神振作的
indefinitely[ɪnˈdefɪnətli]   adv. 不确定地，无限期地
unequalled[ʌnˈiːkwəld]   adj. 无与伦比的；不等同的，不能比拟的
alluring[əˈlʊrɪŋ]   adj. 诱惑的，诱人的；迷人的，吸引人的
souvenir[ˌsuːvəˈnɪr,ˈsuːvənɪr]   n. 纪念品；礼物
certificate[sərˈtɪfɪkət]   n. 证书；文凭，合格证书
elegant[ˈelɪɡənt]   adj. 高雅的，优雅的
fancy[ˈfænsi]   v. 想做；喜爱；自负；想象   n. 幻想；想象力；爱好   adj. 复杂的；昂贵的；精致的，花哨的；想象的

Practical Activities 实践活动

Part A Translate the top 10 world's best selling perfumes brands.

Rank	Brand	Product	Introduction	Translation
1	Chanel		Chanel's colognes are famed because of their durable and pleasing fragrance. Chanel No. 5 is the first scent introduced by the company and it became the reason of boosting the sales	
2	Prada		After Chanel, the only high-quality scent can be found in Prada. They produce fragrances for both men and women but mostly for the female segment	

续表

Rank	Brand	Product	Introduction	Translation
3	Gucci		Gucci's fragrance for men includes Guilty, Oud and Gucci while for women, they have flora, bamboo and others. The bottles are extremely attractive and classy that the customers who visit the outlets can't come out without buying at least one	
4	Hermès		Masculine and feminine fragrances are available. Galop D. Hermes is the most recent perfume for women. Also, unisex perfumes are there including the Garden Collection, Cologne Collection, Eau De Hermes and Voyage De Hermes	
5	Christian Dior		Miss Dior is the first perfume launched by the company in 1947 while the second one Diorama was introduced in 1949. The fragrance 'Poison Girl' is extremely catchy and it's almost impossible to ignore it when it's on the display	
6	Versace		No matter you see Versace online or offline, you won't waste a single second for falling in love with the beautiful and appealing packed bottles. There is a huge variety of perfumes for men and women	
7	Tom Ford		These perfumes are distinguished because of the decent packaging and distinctive fragrance. This luxurious scent generates sex appeal along with eternal stylishness. This premium perfume brand is one of the most wanted colognes worldwide generating high sales and earning revenues from the scent market	

Lesson 25 Selling Cosmetics & Perfume

续表

Rank	Brand	Product	Introduction	Translation
8	Burberry		Since 161 years, this brand has been serving the world with premium products and now has built a strong image in the minds of the buyers. The title of the brand is catchy and unique which is a major reason for its fame	
9	Bulgari		The brand is written as BVLGARI as the logo in the classical Latin alphabet. The fragrances were launched in the early 1990s. Get ready for the new launches and fresh products from this luxurious brand	
10	Armani		Armani is famous for producing ready-to-wear as well but the fragrances they manufacture are up to the mark. This fastest growing fashion brand has won the hearts of millions of its consumers and the market shares are increasing day by day	

Part B Group discussion: do you know any other famous brands?

Lesson 26 Make-up & Hairdressing
化妆及美发

Lead in 课前导入

Thinking and talking:
1. Do you know the skills of putting on various makeup for customers?
2. How about the skills of hairdressing in a hair salon?

Related words:

daytime makeup	日妆	bridal make up	新娘妆
party makeup	宴会妆	makeup artist	化妆师
hair styling	发型设计	hair salon	美发店
hairdresser	美发师		

Lesson 26 Make-up & Hairdressing

Task 1 Makeup Service

任务一 化妆服务

Sample Dialogues 对话范例

Dialogue 1: Day time make-up 化日妆

A: Good morning, Miss Wang. How beautiful you look today!

B: Thank you. I have had make-up.

A: Who taught you to put on make-up?

B: I have studied make-up at a beauty shop.

A: Can you show me the day time make-up?

B: Of course. First, you should always apply basic skin care so that your skin won't be hurt. After that, you can use concealer and foundation to conceal facial defects, improving the look of your skin and even modify your facial features. No matter the shape of your face is square, round or triangle, proper foundation will make you look good.

A: After that, what else should I put on?

B: There follows the part of eye make-up. Firstly trim and draw eyebrow with an eyebrow pencil. Then it's important to wear eye shadows in three similar colors and then wear eye liner and mascara. Your eyelashes are bit long, but not very thick and curly, so I suggest you curl your eyelashes, and then use L'Oreal Paris double extension mascara to make your eyes bigger and more beautiful.

A: What color of eye shadow do you think is the most fitful for me?

B: I think the coral red set is most popular among Chinese girls.

A: How about lips?

B: I think you'd better use blush and lipstick, which can improve your complexion and facial features.

A: Is that all?

B: Well, the most crucial and final work is to set your makeup with loose powder, so that your make will last longer.

A: I see. Thanks a lot. I'll try it.

B: You're welcome.

Dialogue 2: Bridal make up 新娘妆

Makeup artist: Good morning. What can I do for you?

Customer: I'm going to my wedding party tomorrow afternoon. I'm the bride. I want to try my makeup in advance.

Makeup artist: Of course. On the wedding day, the bride should be the most beautiful lady and I'm sure to make you the shining focus that day.

Customer: Thank you. I never wear too much makeup. What color should I use for my lips?

Makeup artist: Sheer pink is lovely and soft for your lips.

Customer: OK.
Makeup artist: And your narrow eyes need definition. Dark brown series eyeshadow works well for defining.
Customer: I wonder if I'll look good wearing this?
Makeup artist: You'll be dazzling.
Customer: I trust you.

Key Words & Phrases   重点词汇

bride[braɪd]   *n.* 新娘
bridegroom[ˈbraɪdgruːm]   *n.* 新郎
best man   伴郎
bridesmaid[ˈbraɪdzmeɪd]   *n.* 女傧相，伴娘
sheer[ʃɪə]   *adj.* 极轻薄的，陡峭的，全然的
narrow[ˈnærəʊ]   *adj.* 狭窄的
definition[ˌdefɪˈnɪʃn]   *n.* 定义（轮廓）清晰度，鲜明度
define[dɪˈfaɪn]   *vt.* 定义；加强，使轮廓分明
dazzle[ˈdæzl]   *v.* （强光等）使目眩，闪光

Practical Activities   实践活动

How to Apply Makeup in 10 Easy Steps? Try to Practice it!

1. Prep Your Skin

Apply a primer or BB cream with a built-in moisturizer and SPF—go for an SPF 15 or higher. These multipurpose products prep, hydrate, and protect your skin in one step.

2. Conceal

Hide under-eye circles, blemishes, and hyperpigmentation with a concealer that matches your skin tone. Use a pointed concealer brush to dab the product on, then blend with your clean finger.

Tip: Professional makeup artists use color-correcting concealers to minimize inflammation and balance skin tone. To hide redness, opt for a green-tinted concealer.

3. Build a Base

Apply foundation to your T-zone and any problem areas. For sheer coverage, use your fingers to blend foundation into your skin; for more coverage, use a foundation brush.

Tip: If you were blessed with an even skin tone, show it off. Use a subtle tinted moisturizer instead of a complexion-covering foundation.

4. Set Your Foundation

Lightly dust loose translucent powder over your foundation to set it and leave a matte finish.

5. Add a Rosy Glow

Use a large soft-bristle brush to apply blush to your cheeks. Start at the apples and then sweep upward toward your temples.

6. Fill in Your Brows

Use an eyebrow pencil or cream to fill in sparse areas and create a well-defined arch. Focus the product on the upper half of your brow line for an instant brow lift.

Tip: If you have brunette or black hair, go a shade or two lighter on your brows. Blondes can emphasize light brows by going a shade or two darker.

7. Accentuate Your Eyes

Swipe a neutral eye shadow across your eyelids and up to your brow bone. For more definition, blend a darker earthy shade, like taupe or charcoal gray, into your eyelid crease.

8. Define Your Eyes

Apply black eyeliner on your upper lash line. Start drawing from the inside corner of the eye to the outer corner. Keep this line thin so that it works well from day to night.

9. Pump Up Your Lashes

Apply two coats of black or brown mascara. Hold the mascara wand, starting at the root of your eyelashes, and move it up in a zigzag motion to ensure every lash is coated.

10. Color Your Lips

Line your lips with a lip liner that coordinates with your lipstick. Then, use a brush to dab the color onto your lips in a downward motion. Choose a medium-hued color for daytime—think rose or coral—and add a layer of shimmery gloss to play up your lips for an evening out.

Tip: Lip color might inadvertently emphasize dry, chapped lips. Before applying the color, buff any dryness with a damp washcloth. Apply lip balm and let it absorb into your lips before adding any color.

Task 2 Hair Styling

任务二 发型设计

Key Vocabulary 重点词汇

Men's Hair   Women's Hair

Hair Styles 发型

Key Sentences 焦点句型

1. I want to make an appointment.
我要预约。
2. What kind of hairstyle do you prefer?
您喜欢种发型?
3. My hair is very thick, so what kind of hairstyle will be easiest to maintain?
我的头发很多,哪种发型比较容易整理?
4. this shoulder-length featherish style will make your face look slimmer.
齐肩的羽毛剪发型可使你的脸型看起来修长些。
5. Just a little trim around the sides.
只要在两侧附近稍微修剪一下。
6. I'd like an Afro-hairstyle.
我想要非洲式的发型。
7. What hairstyle do you feel will look best on me?
你觉得什么样的发型看起来最适合我?
8. I'm afraid I dislike this style.

Lesson 26 Make-up & Hairdressing

恐怕我不喜欢这个发型。

9. Show me some pictures of different styles

给我看一些不同发型的照片。

10. Can you do my hair in this style?

你能帮我做这种发型吗？

11. How would you like set, madam?

夫人，您想做什么样的发型？

12. Would you like to look up the computer to see which one fits you best?

您要不要看电脑，看一看哪一种发型对您最合适？

13. Here are the photos of the latest hair styles, madam.

夫人，给您最新发式照片。

14. you'll look very smart with this hair-do, ma'am.

夫人，您理这个发式会显得很时髦。

15. How would you like your hairstyle today?

您今天想要做个什么发型？

16. Would you keep the same fashion?

您要保持原来的发型吗？

17. What hairstyle do you feel will look best on me?

你觉得什么样的发型看起来最适合我？

18. I like this style.

我喜欢这个发型。

19. Show me some pictures of different styles.

给我看一些不同发型的图片。

20. Please do my hair in your style.

请按你的发型给我做。

Sample Dialogues 对话范例

Dialogue 1: hair style 做发型

A guest Linda walks into the hair salon.

Hairdresser: Good morning, madam.

Linda: Good morning. I would like a shampoo and hair set.

Hairdresser: Yes, madam. What style do you want?

Linda: I'd like to try a new hair style. Could you show me some pictures of hair styles?

Hairdresser: Sure. We have various models: hair bobbed, hair sweptback, chaplet hair style, shoulder-length hair style, hair done in a bun. Please have a look at them, madam.

Linda: Thanks. Please give me the style in this

picture here but make the wave longer. I would like hair spray, please.

Hairdresser: Yes, madam.

Linda: Oh, your hair dryer is too hot. Would you adjust it, please?

Hairdresser: Sorry, madam. I'll adjust it right away. Is that all right now?

Linda: Yes, thanks.

Hairdresser: Please have a look.

Linda: Beautifully done. Please trim my eyebrows and darken them.

Hairdresser: All right, madam. And would you like a manicure?

Linda: Yes. Use a light nail varnish, please.

Dialogue 2: having a haircut 剪发

George: I'd like a haircut, please.

Barber: Would you care for a shave and a shampoo as well?

George: No, thanks. A haircut will be just fine.

Barber: All right. How do you like your hair cut?

George: Don't cut it too short on the sides and the back. Just trim it a little.

Barber: How about on top?

George: You can thin the top out a little, but just a little.

Barber: Very well.

George: Say, my hair is kind of oily, and dandruff bothers me very much. I've tried several shampoos in vain. Can you recommend me something effective?

Barber: Well, have you tried Head And Shoulders? It's supposed to be good for the dandruff.

George: I'll try it.

Barber: And you can try Vidal Sassoon's hair tonic. It's used after you wash your hair. It'll keep your hair clean-looking and oil-free.

George: I'll try that, too. Thank you, barber.

Barber: It's done. That will be thirty-five dollars and fifty cents.

Dialogue 3: having a perm 电烫发

Hairdresser: It's been a long time, Mrs. Lee.

Mrs. Lee: Yes. I went to Hawaii on a vacation with my husband.

Hairdresser: When did you come back? Did you have a good time?

Mrs. Lee: We came back the day before yesterday. I enjoyed myself there very much. The

beach was beautiful. You should go there some day. The sun was lovely, too.

Hairdresser: I will. How would you like to do your hair today? The same style as usual?

Mrs. Lee: I have a special party to attend tonight, and I'd like to change styles...Actually, I'm thinking about a perm. My sister is getting married next month. I think if I have a perm

Lesson 26 Make-up & Hairdressing

now, then it'll look natural by then. What do you say?

Hairdresser: That's true. Here are some samples of hair styles. What do you think about this one?

Mrs. Lee: No, I don't like short hair... I like this one. The wave looks beautiful, and fits my age too.

Hairdresser: Very well. You're not in a hurry, are you?

Mrs. Lee: No. You can take your time... Oh, I also want a manicure while I'm having the perm.

Hairdresser: OK. The manicurist will be right here.

Dialogue 4: going to the beauty salon 到美容院

(A phone call rings in the Angel Beauty Salon...)

Amanda: Hello, is this Angel Beauty Salon?

Beauty Salon: Yes, it is.

Amanda: I want to reserve the time for doing my hair.

Beauty Salon: O.K. Let me check... You can come over at three o'clock.

Amanda: O.K. My name is Amanda Wu.

Beauty Salon: We'll be waiting for you. Bye.

(In the afternoon...)

Hairdresser: Good afternoon, Miss. May I help you?

Amanda: I've made a reservation this morning. You told me come over at three o'clock.

Hairdresser: Your name, please.

Amanda: Amanda Wu.

Hairdresser: Oh, yes. Please come over and sit in this chair.

Amanda: OK.

Hairdresser: How do you want your hair done?

Amanda: I want to cut it shorter and have a permanent wave.

Hairdresser: No problem.

(After the hair service...)

Hairdresser: Do you like your new hair style?

Amanda: Yes, it looks fabulous. By the way, I need a manicure.

Hairdresser: I'll get another hairdresser to manage it.

(After the nail service...)

Manicurist: Well, it's done. Do you like it?

Amanda: Great. How much should I pay?

Manicurist: One hundred and twenty-five dollars, please.

Amanda: Here you are.

Manicurist: Thank you, Miss. Please come again.

Key Words & Phrases 重点词汇

bob [bɒb] *n.* 剪短(头发)

sweptback[ˈsweptˌbæk] *adj.* 向后倾斜的
chaplet[ˈtʃæplɪt] *n.* 花冠
spray[spreɪ] *v.* 喷，洒
adjust[əˈdʒʌst] *v.* 调整
trim[trɪm] *v.* 修剪
manicure[ˈmænɪkjʊə(r)] *n.* 修甲 *v.* 修甲
nail varnish 指甲油
barber[ˈbɑːbə(r)] *n.* 理发师
haircut[ˈheɪkʌt] *n.* 剪头发
shave[ʃeɪv] *n.* 修面
dandruff[ˈdændrʌf] *n.* 头皮屑
in vain 徒然，无效
recommend[ˌrekəˈmend] *v.* 推荐
hair tonic *n.* 生发油/水，护发素

Practical Activities 实践活动

Fill in the blanks with the following sentences and practice the dialogues with partners.

1. I'm sure it'll look quite nice on you, madam.
2. Shall I recommend some new styles, madam?
3. I get tired of the same all the time.
4. Then I'll have a shampoo and set.

Dialogue 1

A: How do you want your hair set, madam? The same as usual?

B: Yes, but you might set the waves a little looser than usual.

A: _____ This one would be fine on you.

B: No, thanks, for a woman of my age.

A: All right, forget it then.

B: Hey! you're pulling my hair.

A: I'm so sorry. I'll loosen this curler a little.

B: That's better.

A: Are you planning to have a manicure while you're under the dryer?

B: No, thank you.

A: 你想做什么样的发型，夫人？和平时一样吗？

B: 是的，不过可以把波浪做得比平时再蓬松一点。

A: 我为你推荐一种新发型行吗，夫人？这种发型很适合你。

B: 不用，谢谢。这不适合我这样的年龄。

A: 好吧，那么别介意。

B: 嗨！你拉了我的头发。

A: 非常抱歉。我将把波浪做得蓬松一点。

Lesson 26 Make-up & Hairdressing

B：那很好。

A：吹风的时候你要做指甲修剪吗？

B：不用，谢谢你。

Dialogue 2

A：I want my hair set.

B：Very well, madam.

A：Have you any pictures of new hairstyles? Id like to try something new.

B：Yes, madam. Here are the latest styles. Look at this one it's very much in vogue now. Your hair is long and such a lovely auburn that it'll look perfect in a knot at the back.

A：But won't it make my face look too round?

B：Oh, no, ＿＿＿＿＿＿＿＿＿＿＿＿＿＿＿＿．

A：All right, do my hair like that, and if it doesn't suit me, you'll simply have to restyle it.

B：Very well, madam.

A：我想做头发。

B：好的，夫人。

A：你们有没有新发型的照片？我想试做一个新的发型。

B：有的，夫人。这些都是最新的发型。请看那一个，现在它十分流行了。您的头发长，而且是好看的红褐色，如果把您的头发在后脑勺挽成一个蝴蝶结，那真是漂亮极了。

A：可是，这样的发型会不会使我的脸看起来太圆了？

B：噢，不会的，我认为这样的发型对您很合适，夫人。

A：好吧，就按这个发型给我做吧。不过，如果它对我不合适，你就得重新设计。

B：好的，夫人。

Dialogue 3

A：How would you like your hair done today? Do you want a perm like you had last time?

B：No, nothing all that fancy today. With summer coming on, my hair is a bit too long now, so I'd better have it bobbed. Could you give it a good trimming all the way round? ＿＿＿＿＿＿＿＿＿＿＿＿＿＿＿＿．

A：Right you are, Mrs. Miller：bob shampoo and set. What shampoo do you prefer?

B：Golden Texture, please.

A：It's always been one of the most popular brands. Now. would you just hold this towel over your eyes while I give you a rinse? That's it. Now for a good rubbing. Ready for the set now? I'll start putting the rollers, clips and pins in. Now, would you come over here under the dryer, please? Here are some magazines for you to look through while you sit there.

A：今天您想要什么发型？像上次一样的电烫吗？

B：不，今天不做那种发型。夏天快来了，我的头发现在有点儿长，所以我最好把头发剪短。你能把我的头发好好修剪一下吗？然后我洗洗头，把头发做成波浪形。

A：好啦，米勒夫人：剪短发，洗头及做波浪发型。您喜欢用哪种洗发水？

B：用金太克丝乔。

A：这种洗发水是最受欢迎的牌子之一。我给您洗头时，请用毛巾捂住双眼好吗？对，就这样。现在好好地擦一擦头发。准备好做发型了吗？我现在开始上发卷，得用夹子和头夹。现在请坐到烘干器罩下面来。这儿有一些杂志可供您翻看。

Dialogue 4

A：Can you do my hair now?

B：Sure thing. Will you please sit over here? Wash and set? By the way. we have a special price on permanents.

A：My permanent is still good for a while. Just a shampoo and set this time. Actually a permanent is simply too expensive.

B：I know what you mean, If I weren't an operator, I couldn't t afford it myself.

A：I'd like to have my hair dyed, too, I'm tired of being the same color.

B：Your hair is so black and beautiful, why change?

A：I want to look new and different. _____ .

B：Why don't you try a wig? Everyone is wearing them now.

A：that's a good idea. They are really getting popular.

A：现在能给我做头发吗？

B：当然。请坐这儿行吗？洗和定型？顺便说一下，这儿烫发有特价。

A：我的烫发暂时还可以。这次只要洗和定型就可以了。事实上是烫发太贵了。

B：我知道你的意思。如果我不是一个美发师，我自己也支付不起。

A：我也想把头发染了，我对一成不变的发色感到厌烦。

B：你的头发如此黑亮漂亮，为什么要改变呢？

A：我希望看起来新颖、别致一点。我厌倦老是同一种发色。

B：为什么你不试试假发？现在每个人都戴的。

A：好主意。假发确实很流行。

Lesson 27 Manicure Service
美甲服务

Lead in 课前导入

Thinking and talking:
1. Do you know the skills of polishing nails?
2. How about the skills of serving customers i in a nail salon?

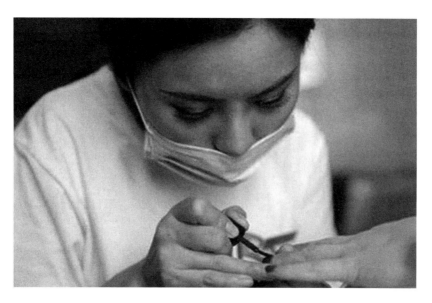

Related words:

1. Nail Polish Products & Tools 美甲产品及美甲工具

manicurist/nail specialist/nail artist	美甲师
nail polish/nail color	指甲油
nail tips	甲片
base coat	底油
quick dry top coat/fast finish top coat	快干亮油
brush on gel/finishing gel	封层胶
manicure	手部护理
pedicure	足部护理
manicure/pedicure package	美甲套餐

basic package/deluxe manicure package	普通/豪华套餐
acrylic nail	水晶甲
French nail	法式水晶甲
gel nail/gel polish	光疗树脂甲
3D nails	3D立体指甲艺术
hand lotion	手霜
hand pillow	手枕
polish remover	洗甲水
clipper	指甲剪
tweezers	镊子
nail dryer	烘甲机
buff	抛光块
colored powders	光疗调色粉
UV lamp	光疗灯
nail glue	甲片胶
nail art brush	指甲彩绘刷
nail extension	指甲加长
foot massager	泡脚机
nail sticker	指甲贴纸
body art tattoo	纹身贴纸
glitter	亮粉
rhinestone	钻饰
manicure bowl	泡手碗
cuticle remover	软化剂

2. 指甲形状分类

square	方形
round	圆形
oval	椭圆形
squoval	方形与椭圆形的结合
almond	杏仁形
coffin/ballerina	两边尖，顶端平，酷似棺板/芭蕾舞鞋的形状
stiletto	剑形/锥形

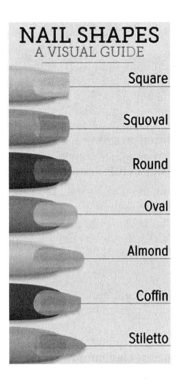

Task 1　Service Process of Nail Care

任务一　美甲服务流程

Key Sentences　焦点句型

1. Guest Greetings　迎宾

（1）欢迎光临！请坐！

Welcome! Please sit down!

（2）（上水。）请喝水！

Please drink some water!

（3）很高兴见到您，请问可以帮您做些什么？

Nice to meet you. What can I help you?

（4）我们可以提供各种专业的美甲服务，请问您今天想享受哪一项？

We can provide a variety of professional nail services. Which one would you want to enjoy today?

（5）这是我们的服务指南，请您欣赏。

This is a guide to our services, please enjoy it.

（6）这些款式都是今年的最新款，不知道有没有您喜欢的呢？

These are this year's latest styles. Are there any styles that you like?

（7）您也可以简单地描绘一下您自己的想法，我们会尽量按照您的意见去做。

You can simply describe your thoughts, and we will try our best to do it according to your opinion.

2. Service Preparation　服务准备

（1）我们先洗一下手好吗？Shall we wash your hands first, OK?

（2）这边请！This way, please!

（3）请先将您的手浸泡二十秒。First, please soak your hands 20 seconds.

(4) 请用洗手液。Please use the hand sanitizer.

(5) 请用毛巾。Please use the towel.

3. Details in the Services 服务细节交流

(1) 你好，请问有你喜欢的款式吗？

Hello, do you have your favorite design in mind?

(2) 请允许我给您简单介绍下好吗？

Please allow me to briefly introduce it to you.

(3) 这个是彩绘指甲。

This is a painted nail.

(4) 这个是纤维甲片。

This is the fiber nail.

(5) 这个是全贴甲片/半贴甲片。

This is all posted nail. /This is a semi-paste nail.

(6) 这个是法式甲片。

This is a French nail.

(7) 这款指甲的售价是 200 元。全程操作时间为一个半小时。

The price of this nail is 200. The whole operation time is one and a half hours.

4. Service Process 服务流程（基础护理）

(1) 消毒双手和工具 sterilize your hands and tools

(2) 涂软皮剂 apply soft skin agent

(3) 修形 modification

(4) 推死皮 push dead skin

(5) 剪死皮 cut dead skin

(6) 粗抛 rough polishing & 细抛 fine polishing

(7) 营养油清洁护理 nutrition oil cleaning care

(8) 手指按摩 finger massage

(9) 清理指芯（收费）clean the finger core（charge）

(10) 涂底油 apply the base oil /base coat

5. See a Visitor Out 送客

(1) 很高兴为您服务，能再耽误您几分钟时间吗？

I am glad to have served you. May I take you a few more minutes' time?

(2) 这里有我们一张服务记录，为了下次能更好地为您提供服务，请您简单的填写一下里面的内容好吗？

Here we have a record of service for better service next time. Would you please simply fill in the form?

(3) 非常感谢！

Many thanks!

(4) 欢迎下次光临！再见！

Lesson 27　Manicure Service

See you next time. Goodbye!

Key Words & Phrases　重点词汇

professional[prəˈfeʃənl]　*adj.* 专业的；职业的
sanitizer[ˈsænitaizə]　*n.* 消毒杀菌剂
fiber[ˈfaɪbɚ]　*n.* 纤维
post[poʊst]　*n.* 岗位；邮件；标杆　*vt.* 张贴；公布；邮递；布置
paste[peɪst]　*v.* (用糨糊)涂，敷；黏合；拼贴
modification[ˌmɑːdɪfɪˈkeɪʃn]　*n.* 修改，修正；改变

Practical Activities　实践活动

Part A Look at the pictures and fill in the English or Chinese translation.

1. _____
手部美甲
包括指甲修理和护理

2. _____
足部美甲
包括脚趾修理和清理

nail polish
nail lacquer
nail enamel

3. _____

nail polish remover

4. _____

gel manicure

5. _____

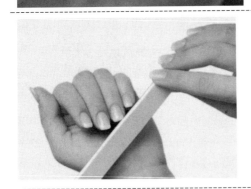

6. _____

美甲图案设计

cutting & filing

7. _____

8. _____

指甲表面抛光

Task 2　Nail Service Dialogues

任务二　美甲服务对话

Key Sentences　焦点句型

1. I want to have a manicure and a pedicure.
   我想做一下手部和脚部指甲的修护。
2. May I also have a French pedicured?

Lesson 27 Manicure Service

我可以也做脚部的法式指甲修护吗？

3. Can you rub in some cuticle cream first and just gently push them back?
你可以先帮我擦一些指甲周围表皮专用的柔肤霜，再轻轻将它们往后推吗？

4. Could you just cut those overgrown cuticles on the sides?
你可不可以只修剪两侧多长出来的表皮？

5. Ouch, it hurts. Could you be a little gentler?
哎唷，好痛。可以稍微轻一点吗？

6. I have a meeting in an hour. Can you put on the quick dry as the topcoat?
我一小时后有一个会议，你可以给我在最上层擦快干油吗？

7. I'd like to do the tip.
我要贴假指甲。

8. How much is it for a designed manicure?
造型式的手部指甲修护要多少钱？

9. Do you offer a special treatment for dry and cracked nails?
你们有针对干裂指甲的特别疗程吗？

10. This raspberry color looks too rich
这个树莓的颜色看起来太浓了。

11. There are a couple of colors that Id like to try
有几种颜色我想试擦看看。

12. Which polish color are you interested in?
您想擦哪种颜色的指甲油？

13. Would you like to have a French manicure?
您想做法式的指甲修护吗？

14. Would you like your nails filed square or round?
您想将指甲修成方形还是圆形？

15. Are you planning to get a manicure later?
您打算以后修指甲吗？

16. Would you like to have a manicure?
您要修指甲吗？

17. What color would you prefer for your nail-polish?
您喜欢什么颜色的指甲油？

Sample Dialogues 对话范例

Dialogue 1

A：Hi, may I help you?
B：I'd like to have a manicure.
A：Sure. Please choose the color you'd like to use.
B：Uh, can I try these two first? I'm not sure which one I'd like to use.
A：Of course. Let me put them on your fingernails.. how is that?
B：I am sorry. I don't like either of them.

A：That's all right. Would you like to try our French manicure? it's very popular now.

B：Really? I'd also like to have a pedicure.

A：嗨，我能为您效劳吗？

B：我想做手部的指甲修护。

A：好的。请选一下您想擦的颜色。

B：嗯，我可以先试一下这两种吗？我不确定我想擦哪种颜色。

A：当然可以。让我把它们涂在你手的指甲上……怎么样？

B：真抱歉，两个我都不喜欢。

A：没关系。您要不要试试我们的法式指甲修护呢？现在很流行哦。

B：真的吗？我同时也想做脚指甲呢。

Dialogue 2

A：Usually, it's ＄60 for a manicure and ＄80 for a pedicure, but with this special discount, the total is only ＄120.

B：That's nice! I'd like to give it a try.

A：All right. Do you want your cuticles cut, too?

A：Would you like the shape square or round.

B：Square, but with rounded edges, please.

A：All right, it's done. Please follow me to the drying section.

B：How long will it take to get them dry?

A：About fifteen minutes, you'll be all set.

B：I see. Thank you.

A：通常修手部的指甲是60美元，修足是80美元，但在这个特惠折扣后，总共只要120美元。

B：很棒啊！那我试试看。

A：好的。您要剪修指甲周围的表皮吗？

B：不要。

A：您喜欢修成方形还是圆形？

B：方形，请将两边稍微磨圆一些。

A：好的。请随我到烘干区。

B：烘干要多久啊？

A：大约15分钟就好了。

B：我知道了，谢谢你。

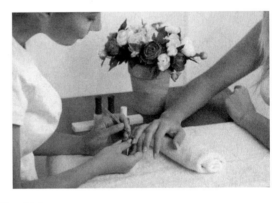

Dialogue 3

A：Excuse me. Can you show me some nail-polish?

B：Ah, yes. We have all kinds of colors for it.

A：that's good. I can choose to my heart's content.

B：Here you are. What's your favorite color?

A：It's hard to say. But white and light purple are very nice colors for nails.

B：Sometimes we don't even understand ourselves. How do you like pink? It matches your-skin.

Lesson 27　Manicure Service

A：Some of my friends like pink nail-polish very much, but I'm not interested in it. Well, just these two, please.

B：Thank you. Anything else?

A：That's all. Maybe I'll come back, but not today.

A：劳驾。可否拿些指甲油给我看看？

B：啊，好的。我们有各种颜色的指甲油。

A：那太好了。我可以尽情地挑选了。

B：给。你最喜欢那种颜色？

A：难说。但对于指甲，白色和淡紫色挺漂亮的。

B：有时，我们连自己都不太了解。你喜欢粉色吗？这跟你的皮肤挺配的。

A：我的一些朋友很喜欢粉色指甲油，但我不感兴趣，嗯，就来这两瓶吧。

B：谢谢。还要其他的吗？

A：就这些了。可能我还会来的，但不是今天。

Dialogue 4

A：I broke one of my nails when I was using my washing machine.

B：Let me look. it's not too bad. fortunately. I think I can take care of it.

A：Thank you. I'm thinking of trying that purple nail polish this time.

B：Good idea. Purple is my favorite color.

A：OK. let's begin.

B：I'm sure I'll be satisfied with it.

A：I hope so.

A：我用洗衣机的时候把一个指甲搞裂了。

B：我来看看，幸好还不太糟。我想我能把它修好。

A：谢谢，我想这次试试那种紫色指甲油。

B：好主意。紫色是我最喜欢的颜色。

A：好的，让我们开始吧。

B：我确信我会满意的。

A：希望如此。

Key Words & Phrases　重点词汇

manicure[ˈmænɪkjʊr]　*n.* 修指甲，美甲，指甲护理

pedicure[ˈpedɪkjʊr]　*n.* 修趾甲术；足部治疗　*vt.* 修脚

cuticle[ˈkjuːtɪkl]　*n.* 角质层；表皮

content[ˈkənˈtent]　*n.* 内容，目录；满足；容量　*adj.* 满意的　*vt.* 满意的

知识拓展

美国美甲行业一瞥

词 汇 表

A

abnormal [æbˈnɔːrml] *adj.* 反常的，不规则的，变态的
absorption [əbˈzɔːrpʃn] *n.* 吸收，全神贯注，专心致志
accelerate [əkˈseləreɪt] *vt.* 使……加快，使……增速 *vi.* 加速，促进，增加
accent [ˈæksent] *n.* 口音，腔调，土音，着重点，强调，重音 *v.* 着重，强调，突出
accentuate [əkˈsentʃueɪt] *v.* 着重，强调，使突出
accommodate [əˈkɒmədeɪt] *vt.* 容纳，使适应，供应，调解
accompanying [əˈkʌmpəniɪŋ] *adj.* 伴随的
acidity [əˈsɪdəti] *n.* 酸度，酸性，酸过多，胃酸过多
acquire [əˈkwaɪər] *v.* （通过努力、能力、行为表现）获得，购得，获得，得到
acronym [ˈækrənɪm] *n.* 首字母缩略词
acrylic acid 丙烯酸
adaptation [ˌædæpˈteɪʃn] *n.* 改编，适应，改编成的作品
address [əˈdres] *n.* 地址，网址，演讲，致辞，称呼，谈吐 *v.* 写姓名地址，演说，向……致辞，处理（问题），称呼，诉说，从事，忙于
adhesive [ədˈhiːsɪv] *n.* 黏合剂，胶带 *adj.* 黏合的，黏性的
adjuvant [ˈædʒəvənt] *adj.* 辅助的 *n.* 佐药，辅助物
administer [ədˈmɪnɪst] *v.* 管理，治理（国家），给予，执行
administration [ədˌmɪnɪˈstreɪʃ(ə)n] *n.* 行政部门
adulterated or misbranded 掺假伪劣或错误标注
adulteration [əˌdʌltəˈreɪʃn] *n.* 掺杂，掺假货
adverse [ədˈvɜːrs] *adj.* 不利的，相反的，敌对的
aerosol [ˈeərəsɒl] *n.* 气溶胶，气雾剂
aerosol products *n.* 气溶胶产品
aggressive [əˈɡresɪv] *adj.* 侵略的，进攻性的，好斗的，有进取心的
alcohol [ˈælkəhɔːl] *n.* 酒精，乙醇
alert [əˈlɜːrt] *adj.* 警惕的，警觉的，机警的，机敏的 *v.* 使警觉，警告，使意识到 *n.* 警戒，警惕；警报
algorithm [ˈælɡərɪðəm] *n.* [计][数] 算法，运算法则
alkalinity [ˌælkəˈlɪnəti] *n.* [化学] 碱度，碱性
alkyl [ˈælkaɪl] *adj.* 烷基的，烃基的 *n.* 烷基，烃基
alkyl sulfate *n.* 烷基硫酸盐
alkylbenzene sulfonate （ABS）烷基苯磺酸盐
all purpose *n.* 通用的，万能的

allegation *n.* 指控，陈述，主张，宣称，陈词，陈述
allergic [əˈlɜːdʒɪk] *adj.* （对……）变态反应的，过敏的，变态反应性的，变应性的，过敏性的，对……十分反感；厌恶
alleviat [əˈliːvieɪt] *vt.* 减轻，缓和
allure [əˈlʊr] *n.* 诱惑力，引诱力，吸引力 *v.* 吸引，引诱
alongside [əlɒŋˈsaɪd] *adv.* 在……的侧面，在……旁边，与……并排 *prep.* 在……旁边，横靠，傍着
alopecian [ˌæləˈpiːʃə] *n.* 脱发症，秃头症
alpha hydroxy acids *n.* 果酸
alter [ˈɔːltər] *vt.* 改变，更改 *vi.* 改变，修改
amino acid [əˈmiːnəʊ ˈæsɪd] *n.* [生化] 氨基酸
ammonium cation *n.* 铵离子
ammonium laureth sulfate *n.* 月桂醇聚醚硫酸酯铵
amplify [ˈæmplɪfaɪ] *vt.* 放大，增强，详述，详加解说
ancestor [ˈænsestər] *n.* 始祖，祖先，被继承人
ancestral [ænˈsestrəl] *adj.* 祖先的，祖传的
ancient [ˈeɪnʃənt] *adj.* 古代的，古老的，过时的，年老的 *n.* 古代人，老人
anhydrous [ænˈhaɪdrəs] *adj.* 无水的
animal origin *n.* 动物来源
anionic [ˌænaɪˈɒnɪk] *adj.* 阴离子的
annual budget *n.* [财政] 年度预算
anti- [ˈænti] *pref.* 反，反对，对立，对立面，防，防止
anti-aging products 抗衰老产品
anti-foaming agent *n.* 防泡剂
antimicrobial [ˌæntimaɪˈkrobiəl] *adj.* 杀菌的，抗菌的 *n.* 杀菌剂，抗菌剂
antistatic agent [助剂] 抗静电剂，静电防止剂
anxiety [æŋˈzaɪəti] *n.* 焦虑，忧虑，担心，害怕，渴望
appearance [əˈpɪərəns] *n.* 外貌，外观，外表，（尤指突然的）到来，起源，出现，首次使用
appliances [əˈplaɪəns] *n.* 家用电器
approach to 接近
approval [əˈpruːvl] *n.* 批准，认可，赞成
aqua [ˈɑːkwə] *n.* 水，溶液，浅绿色 *adj.* 浅绿色的
arbitrary [ˈɑːbɪtrəri] *adj.* 任意的，武断的，专制的
archaeologist [ˌɑːrkiˈɑːlədʒɪst] *n.* 考古学家
archeological [ˌɑːkiəˈlɒdʒikəl] *adj.* 考古学的
aroma [əˈrəʊmə] *n.* 香气，芬芳，芳香
aromatic 芳香的，有香味的
artificial [ˌɑːtɪˈfɪʃl] *adj.* 人造的，仿造的，虚伪的，非原产地的，武断的
aryl [ˈærail; -rɪl] *n.* 芳香基 *adj.* 芳香基的

assessment [ə'sesmənt] n. 看法，评估
association [ə,səusi'eiʃn] n. 协会，社团，联盟
assure [ə'ʃuə] vt. 向……保证，使……确信
attachment [ə'tætʃmənt] n. 附件
attentive [ə'tentɪv] adj. 注意的，体贴的，留心的
attraction [ə'trækʃn] n. 吸引，吸引力；引力；吸引人的事物
audit ['ɔːdɪt] n. 审计，稽核；查账；审查，检查
aura ['ɔːrə] n. 光环，气氛，(中风等的) 预兆，气味
authorized ['ɔːθəraɪzd] adj. 授权的 v. 授权
autonomy [ɔː'tɑːnəmi] n. 自治，自治权
avatar ['ævətɑːr] n. (印度教，佛教) 神的化身，(某种思想或品质) 化身，(网络) 头像，替身

B

backache ['bækeɪk] n. 背痛，腰痛
bacteria [bæk'tɪriə] n. 细菌
ban [bæn] v. 禁止，取缔，(官方) 把 (某人) 逐出某地 n. 禁止，禁令，禁忌，剥夺公民权的判决，诅咒，巴尼 (罗马尼亚货币单位)
barrier ['bæriə(r)] n. 障碍物，屏障，界线 vt. 把……关入栅栏
base notes 尾调
basil/ocimum basilicum 甜罗勒
batch [bætʃ] n. 一批，一炉，一次所制之量 vt. 分批处理
bath product n. 卫浴产品
be aware of 意识到
be commensurate with 与……相应；与……相当
be comprised of 由……组成
be undertaken by 由……承担
beauty parlour 美容院
benzyl paraben 对羟基苯甲酸苄酯
bergamot/Citrus bergamia 佛手柑
betaine ['biːteɪn] n. 甜菜碱
bilayer ['baɪˌleə] n. 双分子层 (膜)
billion ['bɪljən] num. 十亿
binder ['baɪndə(r)] n. [胶黏] 黏合剂，活页夹，装订工，捆缚者，用以绑缚之物
biodegradation ['baɪəu,degrə'deɪʃn] n. [生物] 生物降解，生物降解作用
bioengineer ['baɪəu,endʒɪ'nɪə] n. 生物工程师
black pepper/piper nigrum 黑胡椒
bleed [bliːd] vt. 使出血，榨取 vi. 流血，渗出，悲痛
blemish ['blemɪʃ] n. 斑点，疤痕，瑕疵 v. 破坏……的完美，玷污
blend [blend] v. 使混合，掺和，(和某物) 混合，融合，(使) 调和，协调 n. (不同类

型东西的）混合品，混合物，（不同事物的）和谐结合，融合

bloat [bləʊt] adj. 肿胀的，鼓起的，饮食过度的，胃胀的 v. 使膨胀，肿胀，腌制，溢出 n. 膨胀；过度，过量，（牛、羊等的）胃气胀

blossom 花丛，花簇

blur [blɜːr] n. 模糊不清的事物，模糊的记忆，污迹 v. 使……模糊不清，变模糊，使暗淡，玷污，沾上污迹

blush [blʌʃ] vi. 脸红，感到惭愧 n. 脸红，红色，羞愧 vt. 使成红色

boost [buːst] vt. 促进，增加

bound to 一定会，必然，绑定到

bovine [ˈbəʊvaɪn] adj. 牛的，似牛的，迟钝的 n. 牛科动物

bovine spongiform encephalitis（bse） n. 牛海绵状脑炎

brackets [ˈbrækɪts] n. 支架，括号，圆括号

branched [brɑːntʃ] adj. 分枝的，枝状的，有枝的

branched hydrocarbon chains 支链烃

breakage [ˈbreɪkɪdʒ] n. 破坏，破损，裂口，破损量

brief [briːf] adj. 简短的，简洁的，短暂的，草率的 n. 摘要，简报，概要，诉讼书

bright fuschia 亮玫红

bright red 亮红色；正红

bromide [ˈbrəʊmaɪd] n. 溴化物

bubble bath n. 泡沫浴；泡泡浴

burial [ˈberiəl] n. 埋葬，葬礼，弃绝 adj. 埋葬的

bursting [ˈbɜːstɪŋ] adj. 充满……的，充满（感情或特点）的，急于（做……）的，渴望（做……）的

butylparaben [betilpəˈræben] n. 尼泊金丁酯，对羟基苯甲酸丁酯

by-products n. 副产物

C

calcium paraben 对羟基苯甲酸钙

calendula oil/Calendula officinalis 金盏花浸泡油

carboxylic acid n. [有机化合物] 羧酸

carcinogenic [ˌkɑːrsɪnəˈdʒenɪk] adj. 致癌的，致癌物的

cash-strapped [kæʃ stræpt] adj. 资金短缺的

cassia/Cinnamomum cassia 中国肉桂

castor oil 蓖麻油

category [ˈkætəɡɔːri] n. 种类，分类，范畴

cationic [ˌkætaɪˈɒnɪk] adj. 阳离子的 n. 阳离子

cedar oil 香柏油，雪松油，杉木油，红桧油

cedarwood/Cedrus atlantica 大西洋雪松

celebrities [sɪˈlebrɪtɪz] n.（尤指娱乐界的）名人（celebrity 的名词复数），名流，名声，名誉

celebrity [sə'lebrəti] *n.* 名人
cellular ['seljələr] *adj.* 细胞的，多孔的，由细胞组成的
Center for Drug Evaluation and Research （CDER）药品评价与研究中心
central nervous system tissue 中枢神经系统组织
centralise ['sentrəlaɪz] *vt.* 把……集中起来，形成中心
ceramide ['serəmaid] *n.* 神经酰胺
cereal ['sɪrɪəl] *n.* 谷类，谷物，谷类食品，谷类植物 *adj.* 谷类的，谷类制成的
certifiable [ˌsɜːrtɪ'faɪəbl] *adj.* 可证明的，可确认的，可保证的
certification [ˌsɜːrtɪfɪ'keɪʃn] *n.* 证明，鉴定，出具课程结业证书，颁发证书
certify ['sɜːrtɪfaɪ] *v.* 证明，保证
cetyl ['sitl] *n.* 十六（烷）基，鲸蜡基
cetyl palmitate *n.* 棕榈酸鲸蜡酯
cheek color 腮红
chemical combination ［化学］化合，化合作用
chloride ['klɔːraɪd] *n.* 氯化物
Christianity [ˌkrɪsti'ænəti] *n.* 基督教，基督教徒，基督教教义
circumstantial [ˌsɜːkəm'stænʃl] *adj.* 依照情况的，详细的，详尽的，偶然的
citric acid ['læktɪk 'æsɪd] *n.* 柠檬酸
civilization [ˌsivələ'zeɪʃn] *n.* 文明，文化
clary sage/salvia sclarea 快乐鼠尾草
classic red 复古红
cleanliness ['klenlinəs] *n.* 清洁
clinical ['klɪnɪkl] *adj.* 临床的，诊所的
Clove/Eugenia caryophyllus 丁香
coal-tar ['koʊltɑːr] *n.* 煤焦油
coal-tar dyes *n.* 煤焦油染料
Cocamidopropyl Betaine *n.* 椰油酰胺丙基甜菜碱
Colipa （European Cosmetics Trade Association）欧洲化妆品行业协会
cologne [kə'ləʊn] *n.* 科隆香水，古龙水 Cologne，科隆（德国城市名）
color additives *n.* 颜色添加剂
colorant ['kʌlərənt] *n.* 着色剂
coloring intensity *n.* 颜色强度，彩色亮度
colors exempt from certification *n.* 免检颜色
colourants 着色剂
commercialise [kə'mɜːʃəlaiz] 商业化，使商业化
commit [kə'mɪt] *vt.* 犯罪，把……交托给，指派……作战，使……承担义务，（公开地）表示意见 *vi.* 忠于（某个人、机构等），承诺
compatibility [kəmˌpætə'bɪləti] *n.* 和睦相处，并存，相容，兼容性，相容性
competency ['kɒmpɪtənsi] *n.* 能力，资格
competent authorities 主管部门，主管当局

complementarity [ˌkɑmpləmɛnˈtærəti] n. 互补性，补充，补足
compliance [kəmˈplaɪəns] n. 顺从，服从；符合
composition [ˌkɑːmpəˈzɪʃn] n. 作文，作曲，构成，合成物
compounded [kəmˈpaʊndɪd] adj.［生物］复合的，化合的 v. 混合，组成
comprehensive [ˌkɒmprɪˈhensɪv] adj. 综合的，广泛的，有理解力的 n. 综合学校，专业综合测验
compression [kəmˈpreʃn] n. 压缩，浓缩，压榨，压迫
conceal [kənˈsiːl] v. 隐藏，隐瞒，掩盖
concealer. [kənˈsiːlə] vt. 隐藏，隐瞒，掩盖
concentration [ˌkɒnsnˈtreɪʃn] n. 浓度
concise [kənˈsaɪs] adj. 简明的，简洁的
conditioner [/kənˈdɪʃənə(r)] n. 护发素
conduct [kənˈdʌkt] v. 组织，实施，进行，指挥（音乐），带领，引导，举止，表现，传导（热或电） n. 行为举止，管理（方式），实施（办法），引导
confirmatory [kənˈfɜːmətəri] adj. 证实的，确实的
confuse [kənˈfjuːz] vt. 使混乱，使困惑
conscious [ˈkɑːnʃəs] adj. 意识到的，故意的，神志清醒的
conspicuous [kənˈspɪkjuəs] adj. 显著的，显而易见的
constituent [kənˈstɪtʃuənt] n. 成分，选民，委托人 adj. 构成的，选举的
consumption [kənˈsʌmpʃn] n. 消费；消耗
contact lenses n. 隐形眼镜
contamination [kənˌtæmɪˈneɪʃn] n. 污染，玷污，污染物
contemporary [kənˈtempəreri] adj. 发生（属）于同时期的，当代的 n. 同代人，同龄人，同时期的东西
continent [ˈkɒntɪnənt] n. 大陆，洲，陆地 adj. 自制的，克制的
contrary [ˈkɒntrəri,kənˈtreəri] adj. 与之相异的，相对立的，相反的，（在性质或方向上）截然不同的，完全相反的 n. 相反的事实（或事情、情况）
contribute to 有助于
conversion [kənˈvɜːʃn] n. 转换，变换
conviction [kənˈvɪkʃn] n. 定罪，确信，证明有罪，确信，坚定的信仰
coprecipitation [ˈkəʊpriˌsipiˈteiʃən] n. 共沉淀
correspondence [ˌkɒrɪˈspɒnd(ə)ns] n. 通信，信件，相符，相似
cosmeceutical [ˈkɒzməˈsjuːtikəl] 药用化妆品
cosmetic [kɒzˈmetɪk] adj. 美容的；化妆用的 n. 化妆品；装饰品
Cosmetic Ingredient Review (CIR) n. 化妆品成分评估
cosmetic products notification portal (CPNP) 化妆品备案门户
costly [ˈkɒstli] adj. 花钱多的，昂贵的，价钱高的，引起困难的，造成损失的
cough syrup n. 止咳糖浆，咳嗽糖浆
council [ˈkaʊns(ə)l; -sɪl] n. （市、郡等的）政务委员会，地方议会
counterfeit [ˈkaʊntəfɪt; -fiːt] n. 赝品，伪造品 adj. 假冒的，假装的 v. 仿造，伪装，

假装

cover up （用……）盖在……上（以保护或隐藏），隐藏，遮掩（事实）

criticism ['krɪtɪsɪzəm] n. 批评，批判，责备，指责，评论文章，评论

cross-border ['krɔːsbɔːrdər] adj.（公司）跨国的

crusade [kruːˈseid] n. 改革运动，十字军东侵 vi. 加入十字军，从事改革运动

crystalline ['krɪstəlaɪn] adj. 透明的，水晶般的，水晶制的

curbed [kɜːb] adj. 约束的 v. 抑制

customization ['kʌstəmaɪzeɪʃən] n. 定制，用户化，客制化服务

cuticle ['kjuːtɪkl] n. 角质层，表皮，护膜

D

daisy 雏菊花

dandruff ['dændrəf] n. 头皮屑

date of minimum durability 保质期

dead surface cells 脱落坏死的表面细胞

decelerate [diːˈseləreit] vt. 使减速 vi. 减速，降低速度

deceptive [dɪˈseptɪv] adj. 欺诈的，迷惑的，虚伪的

declaration [dekləˈreɪʃn] n.（纳税品等的）申报，宣布，公告，申诉书

decorative ['dekəreɪtɪv] adj. 装饰性的，装潢用的

dedicated counter 专柜

deep plum 深紫色，深褐紫色

define [dɪˈfaɪn] 解释（词语）的含义，阐明，明确，界定，画出……的线条，描出……的外形，确定……的界线

definition [defɪˈnɪʃ(ə)n] n.（尤指词典里的词或短语的）释义，解释

deforestation [diːˌfɔːrɪˈsteɪʃn] n. 毁林，采伐森林，森林开伐，烧林

deleterious [deləˈtɪəriəs] adj. 有毒的，有害的

delicate 精美的，雅致

demonstrate ['demənstreit] v. 证明，证实，论证，说明，表达，表露，表现，显露，示范，演示

denaturant [diːˈnetʃərənt] n. 变性物质；酒类变性剂

dentifrice ['dentəˌfrɪs] n. 牙膏，牙粉

deodorant [diːˈəʊdərənt] n. 除臭剂，体香剂 adj. 除臭的，防臭的

deodorize [dɪˈəʊdəraɪz] v. 除去……的臭气；给……防臭

Department of Justice 司法部

department store 百货公司，百货商店

depilation [depɪˈleɪʃən] n. 脱毛

derived [dɪˈraɪvd] adj. 导出的，衍生的 v. 衍生出，源于，得到，提取，导出

dermal ['dɜːməl] adj. 真皮的，皮肤的

dermal acute and subchronic studies n. 皮肤急性和亚慢性研究

dermal exposure n. 皮肤接触

dermal sensitization n. 皮肤致敏
desert shrub n. 荒漠灌丛
designate [ˈdezɪgneɪt] vt. 指定，指派，标出 adj. 指定的，选定的
detain [dɪˈteɪn] vt. 拘留，留住，耽搁
detergent [dɪˈtɜːdʒənt] n. 清洁剂，去垢剂
differentiate [ˌdɪfəˈrenʃieɪt] vi. 区分，区别
dihydrate [daɪˈhaɪdreɪt] n. 二水合物，二水物
dilemma [dɪˈlemə] n. 困境，进退两难，两刀论法
diligent [ˈdɪlɪdʒənt] adj. 勤勉的；用功的，费尽心血的
dimension [daɪˈmenʃn] n. 方面，维，尺寸，次元，容积
dimethicone, polydimethylsiloxane n. 聚二甲基硅氧烷
dimethyl sulfuoxide 二甲亚砜
discern [dɪˈsɜːrn] vt. 觉察出，识别，了解，隐约看见 vi. 辨别
discoloration [dɪsˌkʌləˈreɪʃn] n. 变色，污点
discontinuous [ˌdɪskənˈtɪnjuəs] adj. 不连续的，间断的
disinfect [ˌdɪsɪnˈfekt] vt. 将……消毒
disposition [ˌdɪspəˈzɪʃn] n. 处置
dissociate [dɪˈsoʊsieɪt] vt. 游离，使分离，分裂 vi. 游离，分离，分裂
distilled water 蒸馏水
distinctive [dɪˈstɪŋktɪv] adj. 独特的，有特色的，与众不同的
distribution [ˌdɪstrɪˈbjuːʃn] n. 分布，分配，供应
domestic [dəˈmestɪk] n. 国内
dominate [ˈdɒmɪneɪt] v. 支配，控制，左右，影响，在……中具有最重要（或明显）的特色，在……中拥有最重要的位置，俯视，高耸于
dosage [ˈdoʊsɪdʒ] n. 剂量，用量
dose [doʊs] n. 剂量，一剂，一服 vi. 服药 vt. 给药，给……服药
dossiers [ˈdɒsieɪ] n. 档案，卷宗；病历表册
dot-com bubble [dɒt kɒm ˈbʌbl] 互联网泡沫
drab [dræb] adj. 单调的，土褐色的 n. 浅褐色，无生气，邋遢，小额 vt. 使无生气
draw attention to 促使……注意
dye [daɪ] n. 染料，染色 v. 染，把……染上颜色，被染色

E

eco-ethical 生态道德
e-commerce [iːˈkɒmɜrs] n. 电子商务
economic crisis [ˌiːkəˈnɒmɪk ˈkraɪsɪs] 经济危机
effect [ɪˈfekt] n. 影响，效果，作用 vt. 产生，达到目的
efficacy [ˈefɪkəsi] n. 功效，效力
egyptian [iˈdʒɪpʃn] adj. 埃及的，埃及人的，埃及语的 n. 埃及人，古埃及语
elevate [ˈelɪveɪt] vt. 提升，举起，振奋情绪等，提升……的职位

emerge [ɪˈmɜːrdʒ] vi. 浮现，摆脱，暴露
emit radiation　释放辐射
emollient [ɪˈmɒliənt] adj. 使平静的，润肤的，镇痛的　n. 润肤霜
emotion [ɪˈmoʊʃn] n. 情感，情绪
emotional [ɪˈmoʊʃənl] adj. 情绪的，易激动的，感动人的
emphasize [ˈemfəsaɪz] v. 强调，重视，着重，使突出，使明显，重读，强调（词或短语）
empowerment [ɪmˈpaʊərmənt] n. 许可，授权
emulsify [ɪˈmʌlsɪfaɪ] vt. 使……乳化　vi. 乳化
emulsion [ɪˈmʌlʃn] n. [药]乳剂，[物化]乳状液
encapsulate [ɪnˈkæpsjuleɪt] vt. 压缩，将……装入胶囊，将……封进内部，概述
endocrine disruption　内分泌干扰
enhance [ɪnˈhæns] v. 提高，增强，增进
enhancement [ɪnˈhænsmənt] n. 增加，放大
enticing [ɪnˈtaɪsɪŋ] adj. 迷人的，诱人的
enzyme [ˈenzaɪm] n. [生化]酶
epidermal [ˌepəˈdɜːml] adj. [解剖][动]表皮的，外皮的
epidermal growth factor（EGF）表皮生长因子
epidermal permeability barrier　表皮渗透屏障
ergot of rye　n. 麦角黑麦
essential [ɪˈsenʃl] adj. 基本的，必要的，本质的，精华的　n. 本质，要素，要点，必需品
establishment registration　企业注册
ester [ˈestə(r)] n. [有机化合物]酯
estrogen [ˈestrədʒən] n. 雌性激素
ether [ˈiːθər] n. 乙醚
ethics [ˈeθɪks] n. 伦理学，伦理观，道德标准
ethoxylated alcohol salts　n. 乙氧基醇盐
ethylhexyl　n. 乙基己基
ethylhexyl palmitate　n. 棕榈酸乙基己酯
ethylparaben [eθɪlˈpærəben] n. 对羟基苯甲酸乙酯
eucalyptus/eucalypptus radiate　尤加利
European Commission　欧盟委员会
European poison centres　欧洲毒物控制中心
European Union　欧洲联盟
EU's inventory of cosmetic ingredients（cosing）欧盟化妆品成分清单
evening primrose oil/oenothera bienmis　月见草油
evergreen shrub　n. 常绿灌木
evolve [ɪˈvɑːlv] vt. vi. 发展，进化，使逐步形成，推断出
excavate [ˈekskəveɪt] vt. 挖掘，开凿　vi. 发掘，细查
excipient [ɪkˈsɪpiənt] n. [药]赋形剂
exclusive [ɪkˈskluːsɪv] adj. 专用的，专有的，独有的，高档的，豪华的，高级的　n. 独家

新闻，独家专文，独家报道
exclusivity [ˌeksklu:ˈsivəti] n. 排外性，独占权，特有性
exempt [ɪɡˈzempt] adj. 被免除（责任或义务）的，获豁免的 v. 免除，豁免 n. 被免除义务者（尤指被免税者）
exemption [ɪɡˈzempʃn] n. 免除，豁免，免税
exfoliate [eksˈfoʊlieɪt] vi. 片状剥落，鳞片样脱皮 vt. 使片状脱落，使呈鳞片状脱落
exhaustion [ɪɡˈzɔːstʃən] n. 筋疲力尽，疲惫不堪，耗尽，用尽，枯竭
exotic [ɪɡˈzɒtɪk] adj. 异国的，外来的，异国情调的
expand [ɪkˈspænd] v. 扩大，增加，增强（尺码、数量或重要性），扩展，发展（业务），细谈，详述，详细阐明
experiment [ɪkˈsperɪmənt] n. 实验，试验，尝试 v. 实验，尝试
expiration [ˌekspɪˈreɪʃ(ə)n] n. 满期，截止，呼气
expiry date 到期日；有效期限
explicitly [ɪkˈsplɪsɪtli] adv. 明确地，明白地
export [ˈekspɔːt] n. 出口
exposure [ɪkˈspoʊʒər] n. 暴露，曝光，揭露，陈列
exposure to [ɪkˈspəʊʒə(r)tu] 暴露，暴露在，暴露于，受到
extend on 延长，延期，扩大，伸展
extension [ɪkˈstenʃ(ə)n] n. 伸展，扩大，延长，延期，[医] 牵引，电话分机
extensive [ɪkˈstensɪv] adj. 广泛的，大量的，广阔的
external [ɪkˈstɜːrnl] adj. 外部的，表面的，[药] 外用的，外国的，外面的 n. 外部，外观，外面
extraordinary [ɪkˈstrɔːdnri] adj. 非凡的，特别的，离奇的，特派的
extreme [ɪkˈstriːm] adj. 极端的，极度的，偏激的，尽头的 n. 极端，末端，最大程度，极端的事物
eye color 眼妆
eye irritation tests n. 眼刺激性试验
eye makeup remover 眼部卸妆产品
eye shadow 眼影
eyebrow pencil 眉笔
eyeliner [ˈaɪlaɪnər] n. 眼线笔

F

fabric softening n. 植物柔软剂
facial [ˈfeɪʃl] adj. 面部的 n. 面部护理，美容
facial makeup products 面部彩妆
fade [feɪd] v. 褪色，凋谢，逐渐消失，使褪色 n. （电影、电视）淡出，淡入
fair [fer] adj. 公平的，美丽的，白皙的，[气象] 晴朗的 adv. 公平地，直接地，清楚地 vi. 转晴 n. 展览会，市集，美人
Fair Packaging and Labeling Act 正确包装和标识法案

fatty acid   *n.* 脂肪酸
FDA's center for food safety and applied nutrition   *n.* 食品药品管理局食品安全和应用营养中心
FDA's voluntary cosmetic registration program（VCRP）   *n.* FDA化妆品自愿注册计划
Federal Food, Drug, and Cosmetic Act   联邦食品药品化妆品法案
feminine ['femənɪn]   *adj.* 女性的，妇女（似）的，阴性的，娇柔的
feminine   有女性气质的，女子气的
fennel/foeniculum vulgare   茴香
fill in [fɪl ɪn]   填塞，填平（缝隙或孔洞），填，填写（表格等），涂满，填充（图形）
filthy ['fɪlθi]   *adj.* 肮脏的，污秽的
fine line   *n.* 细皱纹
flagship ['flæɡʃɪp]   *n.* 旗舰，一流，佼佼者
flammable ['flæməbl]   *adj.* 易燃的，可燃的，可燃性的
flavor ['fleɪvə]   *n.* 味道
floral   花的，花似的
flourishing ['flʌrɪʃɪŋ]   *adj.* 繁荣的，繁茂的，盛行的
fluffy ['flʌfi]   *adj.* 蓬松的，松软的，毛茸茸的，无内容的
fluidity [fluˈɪdəti]   *n.* 流变性；流质；易变性
foam booster   *n.* 发泡剂
Food and Drug Administration（FDA）   美国食品和药品管理局
foothold   *n.* 据点；立足处
forecast ['fɔːkɑːst]   *n.*（天气、财经等的）预测，预报，预想   *v.* 预报，预测，预示，预言
foreign ['fɒrɪn]   *n.* 国外
formaldehyde [fɔːˈmældɪhaɪd]   *n.* 甲醛
formamide [fəˈmæmaɪd]   *n.* 甲酰胺
formula ['fɔːmjələ]   *n.* 公式，配方，外方，规则，一定的做法
formulation [ˌfɔːmjuˈleɪʃn]   *n.* 制订，规划，（想法或理论的）系统阐述，表达方式，制剂，配方
foundation [faʊnˈdeɪʃn]   *n.* 基础，地基，基金会，根据，创立
fragrance ['freɪɡrəns]   *n.* 芬芳，香味，香气
frankincense/boswellia carteri   乳香
freckle ['frekl]   *n.* 雀斑，斑点
free of charge   免费
Fresh greens   青草香
freshness ['freʃnəs]   *n.* 新，新鲜，精神饱满
friction ['frɪkʃn]   *n.* 摩擦，摩擦力
frown [fraʊn]   *vi.* 皱眉，不同意   *vt.* 皱眉，蹙额   *n.* 皱眉，蹙额
fruity   果香味浓的
fungi ['fʌŋɡi]   *n.* 真菌，菌类，蘑菇（fungus的复数）

G

gamification [ˌgeɪmɪfɪˈkeɪʃn] *n.* 游戏化
garment [ˈgɑːmənt] *n.* 衣服，服装，外表，外观
generally recognized as safe (GRAS) *n.* 通常被认为是安全的物质
genetic mutation *n.* 基因突变
genotoxic [ˌdʒɛnəʊˈtɒksɪk] *adj.* 遗传毒性的
genuine [ˈdʒenjuɪn] *adj.* 真实的，真正的，诚恳的
German chamomile/matricaria recutita 德国洋甘菊
ginger/zingiber officinalis 姜
glycerin [ˈglɪsərɪn] *n.* 甘油
glycerol [ˈglɪsərɪn] *n.* 甘油
glycolic acid [glaɪˈkɑlɪk] *n.* 羟基乙酸
granular [ˈgrænjələ(r)] *adj.* 颗粒的，粒状的
grapefruit/citrus paradisi 葡萄柚
grapeseed oil/vitis vinifera 葡萄籽油
grooming [ˈgruːmɪŋ] *n.* （动物）刷洗，梳毛，梳妆，培养 *v.* （动物）刷洗，梳毛，梳妆，培养
Guangdong Provincial Medical Products Administration 广东省药品监督管理局

H

hair coloring *n.* 染发
hair perming 烫发
hair products 发用化妆品
hairlessness [ˈheələs] *n.* 无毛
harsh [hɑːʃ] *adj.* 严厉的，严酷的，刺耳的，粗糙的，刺目的，丑陋的
hazardous [ˈhæzədəs] *adj.* 有危险的，冒险的，碰运气的
heralded [ˈherəldɪd] *v.* 预示（herald 的过去式和过去分词），宣布（好或重要）
herbal 药草的，草本的
highlight [ˈhaɪlaɪt] *v.* 突出，强调，使醒目，挑染（将部分头发染成浅色） *n.* 最好（或最精彩、最激动人心）的部分，挑染的头发，（图画或照片）强光部分
hormone [ˈhɔːrmoʊn] *n.* [生理] 激素，荷尔蒙
hotel amenities 酒店设施
household cleaning products *n.* 家用清洁产品
hue [hjuː] *n.* 色彩，色度，色调，叫声
human clinical studies *n.* 人体临床研究
human skin tests *n.* 人体皮肤测试
humectant [hjuːˈmektənt] *adj.* 湿润的，湿润剂的 *n.* [助剂] 湿润剂
hydrate [haɪˈdreɪt] *n.* 水合物，水化物 *v.* 补充水分，（使）水合
hydrochloric acid [ˌhaɪdrəˈklɒrɪkˈæsɪd] *n.* 盐酸

hydrogenated [haɪˈdrɑːdʒəneɪtɪd] adj.（油类）氢化的，加氢的
hydrolysis [haɪˈdrɒlɪsɪs] n. 水解作用
hydrolytic [ˌhaɪdrəˈlɪtɪk] adj. 水解的，水解作用的
hydrolyzed [ˈhaɪdrəlaɪzd] adj. 水解的
hydroxypalmitoyl sphinganine 角羟棕榈酰二氢鞘氨醇，神经类鞘脂
hygiene [ˈhaɪdʒiːn] n. 卫生，卫生学，保健法
Hygiene Licence of Imported Cosmetics 进口化妆品卫生许可证

I

identification [aɪˌdentɪfɪˈkeɪʃ(ə)n] n. 鉴定，辨认，确认，确定，身份证明
identity [aɪˈdentəti] n. 身份，同一性，一致，特性，恒等式
illegal [ɪˈliːgl] adj. 非法的，违法的，违反规则的
immersive [ɪˈmɜːrsɪv] adj. 拟真的，沉浸式的，沉浸感的，增加沉浸感的
impact [ˈɪmpækt] n. 巨大影响，强大作用，撞击，冲撞，冲击力 v.（对某事物）有影响，有作用，冲击，撞击
impart [ɪmˈpɑːrt] vt. 给予（尤指抽象事物），传授，告知，透露
import [ɪmˈpɔːt] n. 进口
impression [ɪmˈpreʃn] n. 印象，效果，影响，压痕，印记，感想
in stores 店内
in tandem with 同，同……合作
in turn 反过来，转而
in use 依次，轮流地
inadvertent [ˌɪnədˈvɜːrtnt] adj. 疏忽的，不注意的（副词 inadvertently），无意中做的
indicator [ˈɪndɪkeɪtə(r)] n. 指示信号，标志，迹象，指示器，指针，转向灯，方向灯
indigo [ˈɪndɪɡoʊ] n. 靛蓝，靛蓝染料，靛蓝色，槐蓝属植物 adj. 靛蓝色的
individuality [ˌɪndɪˌvɪdʒuˈæləti] n. 个性，个人，个人特征，个人的嗜好（通常复数）
induce [ɪnˈdjuːs] v. 劝说，诱使，引起，导致，引产，催生
infection [ɪnˈfekʃ(ə)n] n. 传染，感染，传染病，染毒物，影响
infinite [ˈɪnfɪnət] adj. 无限的，无穷的；无数的，极大的
influx [ˈɪnflʌks] n. 流入，汇集，河流的汇集处
ingest [ɪnˈdʒest] vt. 摄取，咽下，吸收，接待
ingredient [ɪnˈɡriːdiənt] n. 成分，（尤指烹饪）原料，（成功的）因素，要素
inhalation [ˌɪnhəˈleɪʃn] n. 吸入；吸入药剂
inherent [ɪnˈherənt] adj. 固有的；内在的；与生俱来的，遗传的
in-house standard 机构内部标准
initial [ɪˈnɪʃəl] adj. 最初的，开始的 n. 首字母
initiate [ɪˈnɪʃieɪt] vt. 开始，创始；发起；使初步了解
innovative [ˈɪnəveɪtɪv] adj. 革新的，创新的；新颖的
insanitary [ɪnˈsænətri] adj. 不卫生的；有害健康的
insect [ˈɪnsekt] n. 昆虫；卑鄙的人

insoluble [ɪnˈsɒljəbl] adj. 不能解决的，不能溶解的，难以解释的
inspection institution　检验机构
institutional [ˌɪnstɪˈtjuːʃənl] adj. 制度的，公共机构的
integral [ˈɪntɪɡrəl] adj. 积分的，完整的，整体的
integrate [ˈɪntɪɡreɪt] vt. 使……完整，使……成整体，求……的积分，表示……的总体 vi. 求积分，取消隔离，成为一体 adj. 整合的，完全的 n. 一体化，集成体
integrity [ɪnˈteɡrəti] n. 完整，正直，诚实，廉正
intended use　预期用途
intensity [ɪnˈtensəti] n. 强度；强烈；紧张
intermediate [ˌɪntərˈmiːdiət] n. 中间中级中间体中间片
interstate commerce　州际贸易
intuitive [ɪnˈtuːɪtɪv] adj. 直觉的，凭直觉获知的
inventory [ˈɪnvəntɔːri] n. 存货，存货清单，详细目录，财产清册
irritating [ˈɪrɪteɪtɪŋ] adj. 刺激的，使愤怒的 v. 刺激，激怒，使烦恼，使发炎，使不适
isobutyl paraben　对羟基苯甲酸异丁酯
isomerized [aɪˈsɒməˌraɪzd] adj. 异构化的
isopropyl alcohol　n. 异丙醇
isopropyl palmitate（IPP）　n. 棕榈酸异丙酯
isopropyl paraben　对羟基苯甲酸异丙酯

J

jasmine/jaminum officinale　茉莉
joint venture　合资公司，合资企业
jurisdiction [ˌdʒʊərɪsˈdɪkʃn] n. 司法权，审判权，管辖权；权限，权力

K

keen [kiːn] adj. 渴望，热切，热衷于，热情的，热心的，喜爱，（对……）着迷，有兴趣
keep up with　跟上
kernel [ˈkɜːrnl] n. 核心，要点，内核，仁；麦粒，谷粒，精髓

L

label instruction　n. 指示标签，标签说明书
label [ˈleɪb(ə)l] n. 标签
lactic acid [ˈlæktɪkˈæsɪd] n. 乳酸
lake [leɪk] n. 湖，深红色颜料，胭脂红，色淀 v.（使）血球溶解
lather [ˈlɑːðə(r)] n. 肥皂泡，激动 vt. 涂以肥皂泡，使紧张，狠狠地打 vi. 起泡沫
laundry [ˈlɔːndri] n. 洗衣店，洗衣房，要洗的衣服，洗熨，洗好的衣服
lauramidopropyl Betaine　n. 月桂酰胺丙基甜菜碱
lavender/lavandula officinalis　薰衣草
leave-on products　n. 驻留类产品

legal entities 法人实体

legal metrology 法制计量

legally ['li:gəli] adv. 合法地；法律上

legislator ['ledʒɪsleɪtə] n. 立法者

lemon/citrus limonum 柠檬

lemongrass/cymbopogon citratus 柠檬香茅

lifespan ['laɪfspæn] n. 寿命；使用期限

light receptor cell n. 光感细胞

limbic ['lɪmbɪk] adj. 边的，缘的

linear alkylate sulfonate (LAS) n. 线性烷基苯磺酸盐

linear structure n. 线性结构

lip balm 润唇膏，护唇膏，唇蜜

lip color 唇妆

lip gloss 唇彩，唇釉

lip liner 唇线笔

lipid ['lɪpɪd] n. 脂质，油脂

lipophilic [lɪpə'fɪlɪk] adj. 亲脂性的，亲脂的

lipstick ['lɪpstɪk] n. 口红，唇膏 vt. 涂口红 vi. 涂口红

low-end market 低端市场

loyalty ['lɔɪəlti] n. 忠诚，忠心，忠实，忠于……感情

lubricant ['lu:brɪkənt] n. 润滑剂；润滑油 adj. 润滑的

luster ['lʌstə] n. 光泽，光彩 vi. 有光泽，发亮 vt. 使有光泽

luxury ['lʌkʃəri] n. 奢侈的享受，奢华，奢侈品

M

Mad Cow Disease 疯牛病

make-up products 化妆产品

malodor [mæl'odə] n. 恶臭，臭气

maltooligosyl-trehalose synthase n. 麦芽寡糖-海藻糖合酶

maltooligosyl-trehalose trehalohydrolase n. 麦芽寡糖-海藻糖水解酶

mandarin/citrus reticulata 橘子

mandate ['mændeɪt] n. 授权，命令，指令，委托管理，受命进行的工作 vt. 授权，托管

mandatory ['mændətəri] adj. 强制的，托管的，命令的

manicure ['mænɪkjʊə(r)] n. 修指甲，美甲，指甲护理

market share n. 市场占有率；市场份额

mascara [mæ'skɑːrə] n. 睫毛膏，染睫毛油 vt. 在……上涂染眉毛油

mask [mæsk] n. 面具，口罩，掩饰 vi. 掩饰，戴面具，化装 vt. 掩饰，戴面具，使模糊

material facts 重要事实

matte 哑光

maximum permissible concentration   n. 最大允许浓度
medication [ˌmedɪˈkeɪʃn]   n. 药，药物
Mediterranean [ˌmedɪtəˈreɪniən]   n. 地中海   adj. 地中海的
melting point   熔点
membranes [ˈmembrens]   n. 细胞膜（membrane 的复数），薄膜，膜皮
mental stress [ˈmentl stres]   精神紧张
methodology [ˌmeθəˈdɑːlədʒi]   n. 方法学，方法论
methylparaben [meθɪlpəˈræben]   n. 对羟基苯甲酸甲酯
microbe [ˈmaɪkroʊb]   n. 细菌，微生物
microbial [maɪˈkroʊbiəl]   adj. 微生物的，由细菌引起的
microbial contamination   微生物污染
microbiologist [ˌmaɪkroʊbaɪˈɑlədʒɪst]   n. 微生物学家
microorganism [ˌmaɪkroʊˈɔːrɡənɪzəm]   n. 微生物，微小动植物
microorganisms [ˌmaɪkrəʊˈɔːɡənɪzmz]   n. 微生物，微生物群，微生物界
middle/heart notes   中调
millennia [mɪˈleniə]   n. 千年期，千周年纪念日
millennial [mɪˈleniəl]   adj. 一千年的，千禧年的
mine [maɪn]   n. 矿，矿藏，矿山，矿井，地雷，水雷   vt. 开采，采掘，在……布雷   vi. 开矿，采矿，埋设地雷   pron. 我的
mineral [ˈmɪnərəl]   n. 矿物，矿泉水，无机物   adj. 矿物的，矿质的
mingle [ˈmɪŋɡl]   vt. 使混合，使相混   vi. 混合起来，相交往
minimalism [ˈmɪnɪməlɪzəm]   极简主义
minimize [ˈmɪnɪmaɪz]   vt. 使减到最少，小看，极度轻视   vi. 最小化
mint-flavored   n. 薄荷味的
miraculous [mɪˈrækjələs]   adj. 不可思议的，奇迹的
miscellaneous [ˌmɪsəˈleɪniəs]   adj. 混杂的，各种各样的；多方面的，多才多艺的
mix [mɪks]   v. （使）混合，配制，参与，交往，交际，混合录音，混成   n. 混合，良莠不齐，混合物
modify [ˈmɒdɪfaɪ]   vt. 修改，修饰，更改   vi. 修改
moisture [ˈmɔɪstʃə(r)]   n. 水分，湿度，潮湿，降雨量
moisturize [ˈmɔɪstʃəraɪz]   vi. 增加水分，变潮湿   vt. 使增加水分，使湿润
moisturizer [ˈmɔɪstʃəraɪzə(r)]   n. 润肤膏
molasses [məˈlæsɪz]   n. 糖蜜，糖浆
molecular weight [化学]   分子量
molecule [ˈmɑːlɪkjuːl]   n. [化学] 分子；微小颗粒，微粒
monomer [ˈmɑnəmər]   n. 单体，单元结构
monosaccharidic [mɒnəˈsækɪrɪdɪk]   n. 单糖，单糖类（最简单的糖类）
monounsaturated [ˌmɒnəʌnˈsætʃəreɪtɪd]   adj. 单不饱和的
moss [mɒs]   n. 苔，藓，地衣
most-favoured-nation (MFN)   最惠国

motivate ['məʊtɪveɪt] v. 刺激，使有动机，激发……的积极性，成为……的动机，给出理由，申请
multi-functional adj. 多功能的
multifunctional [,mʌltɪ'fʌŋkʃənl] adj. 多功能的，起多种作用的
Mummification [,mʌməfəˈkeʃən] n. 木乃伊化
muscle ['mʌsl] n. 肌肉，力量 vt. 加强，使劲搬动，使劲挤出 vi. 使劲行进
muscular aches ['mʌskjələ(r)eɪks] 肌肉疼痛
musk [mʌsk] n. 麝香，能发出麝香的各种各样的植物，香猫

N

Naming Requirements for Cosmetics 化妆品命名要求
nanomaterial 纳米材料
nascent ['neɪsnt] adj. 新兴的，初期的，开始存在的，发生中的
National Medical Products Administration（NMPA） 国家药品监督管理局
natural cosmetics 天然化妆品
natural occurrence 自然发病，天然产状
naturally-occurring 自然发生的
navigation [,nævɪ'geɪʃn] n. 航行，航海
neroli/citrus aurantium bigarade 橙花
nerve [nɜːv] n. 神经，神经质，神经紧张，勇气，气魄 v. 鼓足勇气，振作精神
net quantity 净含量
nominal capacities 额定容量，公称容积
nominal net 净含量
nominal quantities 标称数量
nonetheless [,nʌnðə'les] adv. 尽管如此
non-irritating n. 无刺激性的
nontoxic [nɒn'tɒksɪk] adj. 无毒的
noticeable ['nəʊtɪsəbl] adj. 显而易见的，显著的，值得注意的
notorious [nəʊ'tɔːriəs] adj. 声名狼藉的，臭名昭著的
nude 裸色
numerical count 数码，数字码

O

obligations [,ɒblɪ'geɪʃn] n. 义务，职责，债务
obtain [əb'teɪn] v.（尤指经努力）获得，赢得，存在，流行，沿袭
occident ['ɒksədənt] n. 西方，欧美国家
occlusive [ə'kluːsɪv] adj. 咬合的，闭塞的 n. 闭塞音
occurrence [ə'kʌrəns] n. 发生的事情，存在的事物，发生，出现，存在
octyl palmitate n. 棕榈酸辛酯
ocular ['ɒkjələ(r)] adj. 眼睛的，视觉的，目击的 n. 目镜

odor ['əʊdə]　*n*. 气味；名声
odorless ['əʊdəlɪs]　*adj*. 没有气味的
offspring ['ɔːfsprɪŋ]　*n*. 后代，子孙，产物
olfactory [ɒl'fæktəri]　*adj*. 嗅觉的
omit [ə'mɪt]　*vt*. 省略，遗漏，删除，疏忽
on numerous occasions　无数次
opacifier [o'pæsəˌfaɪɚ]　*n*. 遮光剂，[化工] 不透明剂
oral care agent　*n*. 口腔护理剂
oral hygiene product　口腔卫生产品
orange red　橘红色
orange sweet/citrus Sinensis　甜橙
organ ['ɔːrɡən]　*n*. 器官，机构，风琴，管风琴，嗓音
organic [ɔːr'ɡænɪk]　*adj*. 有机的；组织的；器官的；根本的
organic groups　*n*. 有机基团
overestimate [ˌəʊvər'estɪmeɪt]　*v*. 高估　*n*. 过高的评估
over-the-counter(otc) [ˌoʊvərðə'kaʊntər]　*adj*. 非处方的
oxidation [ˌɑːksɪ'deɪʃn]　*n*. 氧化
ozone-depleting　消耗臭氧层的

P

palette ['pælət]　*n*. 调色板；颜料
Palmitate ['pælmɪteɪt]　*n*. 棕榈酸酯，棕榈酸盐
palmitic acid　*n*. 棕榈酸
panel ['pæn(ə)l]　*n*. 仪表板；嵌板
paprika [pə'priːkə]　*n*. 辣椒粉红辣椒
paraben　对羟基苯甲酸酯；防腐剂
parabens [pə'ræbenz]　*n*. 对羟苯甲酸酯，对羟基苯甲酸酯类，尼泊金类
para-hydroxybenzoic acid(phba)　*n*. 对羟基苯甲酸（phba）
pedicure ['pedɪkjʊə(r)]　*n*. 修趾甲术，足部治疗
pentyl paraben　对羟基苯甲酸戊酯
peppermint/mentha piperita　胡椒薄荷
perceive [pə'siːv]　*v*. 注意到，意识到，察觉到，将……理解为，将……视为，认为
perception [pər'sepʃn]　*n*. 认识能力，知觉，感觉，洞察力，看法，获取
perfume ['pɜːfjuːm]　*n*. 香水，香味　*vt*. 使……带香味　*vi*. 散发香气
periodic [ˌpɪəri'ɒdɪk]　*adj*. 周期的，定期的
periodically [ˌpɪəri'ɒdɪkəli]　*adv*. 定期，周期性
perm [pɜːm]　*n*. 电烫发
permanent wave　烫发
permeate ['pɜːmieɪt]　*vt*. & *vi*. 弥漫，遍布，散布，渗入，渗透
permit [pər'mɪt]　*vi*. 许可，允许　*vt*. 许可，允许　*n*. 许可证，执照

person to person　面对面
personal care products　*n.* 个人护理用品
personalization [ˈpɜːsənəlaizeiʃn]　*n.* 个性化
pertain [pərˈteɪn]　*v.* 适合；关于；适用；从属，归属；生效，存在
pertaining [pɜːˈteɪnɪŋ]　*adj.* 附属的，与……有关的　*n.* 关于（pertain 的 ing 形式）
petroleum [pəˈtroʊliəm]　*n.* 石油
pH adjusters　pH 值调节剂
pharaoh [ˈferoʊ]　*n.* 法老，暴君
pharmaceutical [ˌfɑːrməˈsuːtɪkl]　*adj.* 制药（学）的　*n.* 药物
phase [feɪz]　*n.* 月相，时期，阶段　*vt.* 分阶段进行，使定相
phenyl paraben　对苯二甲酸苯酯
pheromones [ˈferəʊməʊnz]　*n.* 外激素，信息素
photo-contact allergenicity　*n.* 光敏性
photoirritation　*n.* 光刺激
phototoxicity [ˌfəʊtəʊtɒkˈsɪsəti]　*n.* 光毒性
phytosphingosines [faɪtoʊˈsfɪŋɡəsin]　*n.* 植物鞘氨醇
pigment [ˈpɪɡmənt]　*n.* 色素，颜料　*vt.* 给……着色　*vi.* 呈现颜色
pimple removal　祛斑
pine/pinus sylvestris　欧洲赤松
pleasantly [ˈplezntli]　*adv.* 愉快地，快活地，和气地，和蔼地
pledge [pledʒ]　*n.* 保证，誓言，抵押，抵押品，典当物　*vt.* 保证，许诺，用……抵押，举杯祝……健康
plump [plʌmp]　*adj.* 饱满的；胖乎乎的　*v.* （使）饱满而柔软；变圆，长胖；重重地放下　*adv.* 突然（或重重）坠地；直接地　*n.* 突然前冲；重重坠落
Poison Prevention Packaging Act　防止有毒物包装法案
polish [ˈpɒlɪʃ]　*n.* 光泽；上光剂；优雅；精良
pollutant [pəˈluːtənt]　*n.* 污染物
polymer [ˈpɑːlɪmər]　*n.* [高分子] 聚合物
pore [pɔːr]　*n.* （皮肤上的）毛孔，黑头，（植物的）气孔，孔隙
postoperative [ˌpəʊstˈɒpərətɪv]　*adj.* 手术后的
potassium ethylparaben　羟苯乙酯钾
potassium methylparaben　对羟基苯甲酸甲酯钾
potassium paraben　对羟基苯甲酸钾
potent [ˈpoʊtnt]　*adj.* 有效的，强有力的，有权势的，有说服力的
potential [pəˈtenʃl]　*adj.* 潜在的，可能的　*n.* 潜能，可能性，电势
poultry [ˈpəʊltri]　*n.* 家禽，家禽肉
powder [ˈpaʊdə(r)]　*n.* 粉末，细面，扑面粉，美容粉　*v.* 抹粉
precious [ˈpreʃəs]　*adj.* 宝贵的，珍贵的，矫揉造作的
precise [prɪˈsaɪs]　*adj.* 准确的，确切的，精确的，细致的，精细的，认真的，一丝不苟的
predict [prɪˈdɪkt]　*v.* 预言，预告

predominance [prɪˈdɒmɪnəns]　n. 优势，卓越

prefix [ˈpriːfɪks]　n. 前缀　vt. 加前缀，将某事物加在前面

pregnant [ˈpregnənt]　adj. 怀孕的，富有意义的

prehistoric [ˌpriːhɪˈstɔːrɪk]　adj. 史前的，陈旧的

premarket [ˌpriːˈmɑːkɪt]　adj. 上市（销售）前的

premium [ˈpriːmiəm]　n. 额外费用，奖金，保险费，（商）溢价　adj. 高价的，优质的

preservative [prɪˈzɜːrvətɪv]　n. 防腐剂，预防法，防护层　adj. 防腐的，有保存力的，有保护性的

press release　新闻稿

preventative [prɪˈventətɪv]　adj. 预防性的

primarily [praɪˈmerəli]　adv. 首先，主要地，根本上

primitive [ˈprɪmətɪv]　adj. 原始的，远古的；简单的，粗糙的　n. 原始人

Principal Display Panel（PDP）　主要展示版面

prior to　在……之前

prioritise [praɪˈɒrə,taɪz]　vt. 给予……优先权，按优先顺序处理　vi. 把事情按优先顺序排好

prominent [ˈprɒmɪnənt]　adj. 突出的，显著的，杰出的，卓越的

propylparaben [ˌprəʊpɪlˈpɑːrəben]　n. 尼泊金丙酯，对羟基苯甲酸丙酯

pros and cons　利弊

prostitute [ˈprɒstɪtjuːt]　n. 卖淫者，娼妓，妓女，男妓

provision [prəˈvɪʒn]　n. 规定，条款，准备，供应品　vt. 供给……食物及必需品

psyche [ˈsaɪki]　n. 灵魂，心智

psychology [saɪˈkɒlədʒi]　n. 心理学，心理，心理特征，心理影响

Puerto Rico [ˈpwɜːtəʊˈriːkəʊ]　波多黎各

Purchase [ˈpɜːtʃəs]　n. 购买，采购，购买的东西，购买项目，握紧，抓牢，蹬稳　v. 买，购买，采购

purified water　净化水

putrid [ˈpjuːtrɪd]　adj. 腐败的，腐烂的

Q

qualification [ˌkwɒlɪfɪˈkeɪʃ(ə)n]　n.（通过考试或学习课程取得的）资格

quantitative risk assessment（qra）　定量风险评价

quaternary ammonium compoundsn　季铵化合物

R

radiation [ˌreɪdiˈeɪʃn]　n. 辐射，发光，放射物

rash [ræʃ]　n. 皮疹

realisation [ˌriːlaɪˈzeɪʃən]　n. 实现，完成

reassurance [ˌriːəˈʃʊərəns] n. 再保证，再安慰
recession [rɪˈseʃn] n. 经济衰退，经济萎缩，退后，撤回
record-filing 备案
record-keeping certificate 备案证书
reflection [rɪˈflekʃn] n. 反射，沉思，映象
reflex [ˈriːfleks] n. 反射动作，本能反应，反射作用
refreshing [rɪˈfreʃɪŋ] adj. 令人耳目一新的，别具一格的，使人精力充沛的，使人凉爽的 v. 使恢复精力，使凉爽，重新斟满，提醒，提示，使想起
registered [ˈredʒɪstəd] adj. 注册的，记名的，登记过的
regularity [ˌregjuˈlærəti] n. 规律性，经常性，匀称，端正，有规则的分布，有规则的东西，有规律的事物
regulation [ˌregjuˈleɪʃn] n. 管理；规则；校准 adj. 规定的，平常的
Regulations on Cosmetics Hygiene Supervision 化妆品卫生监督条例
reinvent [ˌriːɪnˈvent] vt. 重新使用，彻底改造，重复发明
rejection [rɪˈdʒekʃn] n. 抛弃，拒绝，被抛弃的东西
relevant [ˈreləvənt] adj. 相关的，切题的，中肯的，有重大关系的，有意义的，目的明确的
relic [ˈrelɪk] n. 遗迹，遗物，废墟，纪念物
relief [rɪˈliːf] n.（不快过后的）宽慰，轻松，解脱，（焦虑、痛苦等）减轻，消除，缓和
remain relevant 与时俱进
remover 卸妆产品
Renaissance [rɪˈneɪsns] 文艺复兴
renewal [rɪˈnjuːəl] n. 更新，恢复，复兴，补充，革新，续借，重申
repeating units n. 重复单元，重复单位，重复链段
reputation [ˌrepjuˈteɪʃn] n. 名声，名誉，声望
requisite [ˈrekwɪzɪt] adj. 必备的，必不可少的，需要的
resolve to 决心，解析至，解决
restlessness [ˈrestləsnəs] n. 不安定，烦躁不安，辗转不安
restraining order 禁令
restriction [rɪˈstrɪkʃn] n. 限制，约束，束缚
retail [ˈriːteɪl] n. 零售 v. 零售，（以某价格）零售 adv. 以零售方式 adj. 零售的
retain [rɪˈteɪn] vt. 保持，雇，记住
reversible [rɪˈvɜːsəbl] adj. 可逆的；可撤消的；可反转的
review [rɪˈvjuː] n. 回顾，复习，评论，检讨，检阅 vt. 回顾，检查，复审 vi. 回顾，复习功课，写评论
revision [rɪˈvɪʒn] n. 修正，复习，修订本
revolutionary [ˌrevəˈluːʃəneri] adj. 革命的，旋转的，大变革的 n. 革命者
rinse away 冲洗掉

rinse-off products　n. 淋洗类产品
ritual ['rɪtʃuəl]　n. 仪式，惯例，礼制　adj. 仪式的，例行的，礼节性的
robust [rəʊˈbʌst]　adj. 强健的，健康的，粗野的，粗鲁的
rose Geranium/pelargonium roseum　玫瑰天竺葵
rose pink　玫瑰色
rose/rosa damascena　玫瑰（大马士革）
rosemary/rosmarinus officinalis　迷迭香
rubbed [rʌbd]　v. 擦（rub 的过去式和过去分词），摩擦，搓
rumor ['ruːmə]　n. 谣言，传闻　vt. 谣传，传说
rumoured ['ruːməd]　adj. 谣传的，传说的，风

S

safety ['seɪfti]　n. 安全
Safety and Technical Standards for Cosmetics　化妆品安全技术标准
Safety Approval Committee of the ministry of health　卫生部化妆品卫生安全性评审委员会
saffron ['sæfrən]　n. 藏红花，橙黄色　adj. 藏红花色的，橘黄色的
salon [səˈlɑːn]　n. 沙龙，客厅，画廊，美术展览馆
sandalwood/santalum album　檀香
saturated ['sætʃəreɪtɪd]　adj. 饱和的，渗透的，深颜色的
scarcity ['skersəti]　n. 不足，缺乏
scentless ['sentləs]　adj. 无气味的，遗臭已消失的
sceptical ['skeptɪkl]　adj. 怀疑论的，怀疑的
scrutinise ['skruːtɪnaɪz]　vi. 作仔细检查；细致观察　vt. 细看；仔细观察或检查；核对
sealed [siːld]　adj. 密封的，未知的
sebum ['siːbəm]　n. 皮脂；牛羊脂
sector ['sektə(r)]　n. 部门，扇形，扇区，象限仪，函数尺　vt. 把……分成扇形
seizure ['siːʒə(r)]　n. 没收，夺取，捕获
self-esteem [ˌself ɪˈstiːm]　n. 自尊，自负，自大
semi-solid　n. 半固体的
sensitization [ˌsensətaɪˈzeɪʃn]　n. 敏化作用，促进感受性，感光度之增强
separate ['seprət]　adj. 分开的，单独的，不同的，各自的，不受影响的　v.（使）分离，分开，隔开，分手，（使）分居，（使）区别（于）　n. 可搭配穿着的单件衣服，抽印本，独立音响设备，土壤划分
serious undesirable effects（SUE）　严重不良影响
set forth　陈述，提出，出发，陈列，宣布
shaft [ʃæft]　n. 竖井，通风井，（电梯的）升降机井，杆，柄　vt. 欺骗，苛待
shampoo [ʃæmˈpuː]　n. 洗发，洗发精　vt. 洗发
shaving ['ʃeɪvɪŋ]　v. 修面，剃（shave 的现在分词）　n. 刨花，刮胡子，削

shaving cream   n. 刮胡膏，剃须膏
shaving product   剃须产品
sheer coral   珊瑚红
shelf-life ['ʃelf laɪf]   n. 保质期，货架期，货架寿命，保存期限，储存期
shininess ['ʃaɪnɪnɪs]   n. 光泽度，反射，反光度，光滑度，亮度
shrub [ʃrʌb]   n. 灌木；灌木丛
side-effects   副作用
signal ['sɪgnəl]   n. 信号，暗号，导火线   vt. 标志，用信号通知   adj. 显著的，作为信号的   vi. 发信号
significant [sig'nifikənt]   adj. 有重大意义的，显著的，有某种意义的，别有含义的，意味深长的
silicon ['sɪlɪkən]   n. 硅，硅元素
simmondsia chinensis (jojoba) seed oil   n. 荷荷巴油
simultaneously [ˌsɪməl'teɪniəsli]   adv. 同时，急切地
since then   从那时起
skin irritation   n. 皮肤刺激性
skin protectant   n. 护肤剂
skin texture and tone   皮肤质地和色调
Skincare products   护肤产品
slash [slæʃ]   vt. 猛砍，鞭打
smell [smel]   n. 气味，嗅觉，臭味   v. 嗅，闻，有……气味，察觉到，发出……的气味
SMEs abbr   中小企业（small and medium-size enterprise）
snag [snæg]   n. 障碍，意外障碍，突出物   vt. 抓住机会，造成阻碍，清除障碍物   vi. 被绊住，形成障碍
social status   社会地位
sodium ethylparaben   羟苯乙酯钠
sodium laureth sulfate   n. 月桂醇聚醚硫酸酯钠
sodium lauryl sulfate   n. 月桂醇硫酸酯钠
sodium methylparaben   羟苯甲酸钠甲酯
sodium paraben   对羟基苯甲酸钠
sodium salt   钠盐
soften ['sɒfn]   vt. 使温和，使缓和，使变柔软   vi. 减轻，变柔和，变柔软
softness ['sɔːftnəs]   n. 温柔，柔和
soil [sɔɪl]   n. 土地，土壤，泥土   vt. 弄脏，污辱   vi. 变脏
solely ['səʊlli]   adv. 单独地，仅仅
solubilize ['sɒljəbəˌlaɪz]   vt. 使溶解，使增溶   vi. 溶解
sore [sɔː(r)]   adj.（发炎）疼痛的，酸痛的，气恼，愤慨，愤愤不平   n. 痛处，伤处
sour ['saʊər]   adj. 酸的，发酵的，刺耳的，酸臭的，讨厌的   vi. 发酵，变酸，厌烦

soybean ['sɔɪˌbin] *n*. 大豆，黄豆
specialty store　专卖店
specification [ˌspesɪfɪ'keɪʃn] *n*. 规格，说明书，详述
spectrum ['spektrəm] *n*. 系列，幅度，范围，光谱
sphere [sfɪr] *n*. 范围，球体　*vt*. 包围，放入球内，使……成球形　*adj*. 球体的
sphingomyelin [ˌsfɪŋəʊ'maɪəlɪn] *n*. （神经）鞘磷脂
spinal cord　*n*. 脊髓
spoil [spɔɪl] *vt*. 溺爱，糟蹋，掠夺　*vi*. 掠夺，变坏，腐败　*n*. 次品，奖品
spoilage ['spɔɪlɪdʒ] *n*. 损坏，糟蹋，掠夺，损坏物
spray [spreɪ] *n*. 喷雾，喷雾剂，喷雾器，水沫　*v*. 喷射
spreadable ['spredəbl] *adj*. 容易被涂开的
sprinkle ['sprɪŋkl] *v*. 撒，洒
spritz [sprɪts] *n*. 喷，细的喷流　*v*. 喷
stakeholders ['steɪkhəʊldə(r)] *n*. 利益相关者；赌金保管者
Standardisation Administration of China（SAC）　中国标准化管理委员会
Standardisation Law of the People's Republic of China　中华人民共和国标准化法
starch [stɑːrtʃ] *n*. 淀粉；刻板，生硬　*vt*. 给……上浆
State Administration for Market Regulation（SAMR）　国家市场监督管理总局
state archival filing　国家备案
static ['stætɪk] *adj*. 静态的，静电的，静力的　*n*. 静电，静电干扰
statistical [stə'tɪstɪkl] *adj*. 统计数据的，统计学的
status. ['steɪtəs] *n*. 地位
stipulate ['stɪpjuleɪt] *v*. 规定，明确要求
straight chain　*n*. 直链
straight color　*n*. 纯品色
strain ['streɪn] *v*. 损伤，拉伤，扭伤，尽力，竭力，使劲，过度使用，使不堪承受
stratum corneum ['kɔːnɪəm] *n*. 角质层，角层，角化层
strict [strɪkt] *adj*. 严格的，绝对的，精确的，详细的
strive [straɪv] *vi*. 努力；奋斗；抗争
subchronic [sʌbk'rɒnɪk] *adj*. 亚慢性的
subject to　服从，受制于
submission [səb'mɪʃn] *n*. 投降，提交（物），服从
submit [səb'mɪt] *vt*. 使服从，主张，呈递，提交
subscription [səb'skrɪpʃn] *n*. 捐献，订阅，订金，签署
substance ['sʌbstəns] *n*. 物质，实质，资产，主旨
substantiate [səb'stænʃɪeɪt] *vt*. 证实，实体化
substitute ['sʌbstɪtjuːt] *n*. 代替者，代替物，代用品，替补（运动员）　*v*. （以……）代替，取代

substratum ['sʌbstreɪtəm] n. 基础，根据，下层
sudsing 起泡
sugar cane [ʃʊɡɚken] n. 甘蔗
sulfate ['sʌlfeɪt] n. 硫酸盐 vt. 使成硫酸盐
sulfonate ['sʌlfəneɪt] n. 磺化，磺酸盐 v. 使……磺化
sulfuric acid [sʌl'fjʊrɪk'æsɪd] n. 硫酸
sunscreen ['sʌnskriːn] n. （防晒油中的）遮光剂；防晒霜
suntan ['sʌntæn] n. 晒黑；土黄色军服；棕色
suntan product 防晒产品
surfactant [sɜː'fæktənt] n. 表面活性剂 adj. 表面活性剂的
surveillance [sɜː'veɪləns] n. 监督，监视
survey ['sɜːrveɪ] n. 调查，测量，审视 vt. 调查，勘测，俯瞰 vi. 测量土地
susceptible [sə'septəbl] adj. 易受影响的，易感动的，容许……的 n. 易得病的人
suspend [sə'spend] vt. 延缓，推迟，使暂停，使悬浮 vi. 悬浮，禁赛
sweet almond oil/prunus amygdalus var dulcis 甜杏仁油
synonymous [sɪ'nɑːnɪməs] adj. 同义的，同义词的，同义突变的
synthetic [sɪn'θetɪk] adj. 综合的，合成的，人造的 n. 合成物
synthetic flavoring substance 合成调味物质
synthetic-organic color n. 有机合成的颜色

T

tailor-made [ˌteɪlər'meɪd] adj. 特制的，裁缝制的
take into consideration 考虑到……
tangible [juː'bɪkwɪtəs] adj. 普遍存在的，无所不在的
tariff ['tærɪf] n. 关税，（旅馆、饭店或服务公司的）价目表，收费表，量刑标准
tattoo [tæ'tuː] n. 纹身 v. （在皮肤上）刺图案
tea tree/melaleuca alternifolia 茶树
temporary ['tempəreri] adj. 暂时的，临时的 n. 临时工，临时雇员
tense [tens] adj. 神经紧张的，担心的，不能松弛的，令人紧张的，绷紧的，不松弛的
territory ['terətɔːri] n. 领土，领域，范围，地域，版图
texture ['tekstʃə(r)] n. 质地，手感，口感，（音乐或文学的）谐和统一感，神韵
therapeutic [ˌθerə'pjuːtɪk] adj. 治疗的，治疗学的，有益于健康的
thorough ['θʌrə] adj. 彻底的，十分的，周密的
thyme/thymus vulgaris 百里香
tint [tɪnt] n. 色彩，浅色 vt. 染（发），给……着色
tissue ['tɪʃuː] n. 纸巾，薄纱，一套 vt. 饰以薄纱，用化妆纸揩去
tissues for transplantation 组织移植
titanium dioxide 二氧化钛

toiletry [ˈtɔɪlɪtrɪ] *n.* 化妆品，化妆用具

tomb [tuːm] *n.* 坟墓，死亡 *vt.* 埋葬

tone [toʊn] *n.* 语气，色调，音调，音色 *vt.* 增强，用某种调子说 *vi.* 颜色调和，呈现悦目色调

toothpaste [ˈtuːθpeɪst] *n.* 牙膏

top/head notes 前调

tout [taʊt] *vt.* 兜售；招徕

toxicity [tɒkˈsɪsəti] *n.* 毒性

toxicological [ˌtɑksəkəˈlɑdʒɪkəl] *n.* 毒理学

toxin [ˈtɑːksɪn] *n.* 毒素，毒质

trade secret *n.* 商业秘密，行业秘密

transformation [ˌtrænsfərˈmeɪʃn] *n.* 转化，转换，改革，变形

transmit [trænzˈmɪt] *vt.* 传输，传播，发射，传达，遗传 *vi.* 传输，发射信号

treatment [ˈtriːtmənt] *n.* 治疗，疗法，诊治，对待，待遇，处理，讨论，论述

trehalase [trɪˈhɑlez] *n.* 海藻糖酶

trehalose [ˈtriːhələʊs] *n.* 海藻糖

trigger [ˈtrɪɡər] *n.* 扳机，起因，引起反应的事，触发器，引爆装置 *v.* 触发，引起，开动（装置）

trillion [ˈtrɪljən] *num.* 万亿

tumor [ˈtjʊmɚ] *n.* 肿瘤，肿块，赘生物

turmeric [ˈtɜːrmərɪk] *n.* 姜黄，姜黄根，姜黄根粉，郁金根粉

tutorial [tjuːˈtɔːriəl] *n.* （大学导师的）个别辅导时间，辅导课，教程，辅导材料，使用说明书 *adj.* 导师的，私人教师的，辅导的

U

ubiquitous [juːˈbɪkwɪtəs] *adj.* 普遍存在的，无所不在的

UI（user interface） 用户界面

umbrella body 协调组织

under certain circumstances *adj.* 有时，在某种情况下（在某种状况下）

unduly [ʌnˈdjuːli] *adv.* 过度地，不适当地，不正当地

uniformity [ˌjuːnɪˈfɔːrməti] *n.* 均匀性，一致，同样

universalise 使一般化，使普遍化，通用化

unpleasant [ʌnˈpleznt] *adj.* 讨厌的，使人不愉快的

unreactive [ˈʌnriˈæktɪv] *adj.* 不起化学反应的，化学上惰性的

unscented [ʌnˈsentɪd] *adj.* 无香味的，无气味的

unstable [ʌnˈsteɪbl] *adj.* 不稳定的，动荡的，易变的

unsubstantiated [ˌʌnsəbˈstænʃieɪtɪd] *adj.* 未经证实的，无事实根据的

upsell 追加销售

USP　独特的销售主张

V

vaccine ['væksiːn]　n. 疫苗，牛痘苗

vandalism ['vændəlɪzəm]　n. 故意破坏他人（或公共）财物罪，恣意破坏他人（或公共）财产行为

vapor ['veɪpə(r)]　n. 蒸汽，水汽　v. 蒸发

vendors ['vendə(r)]　n. 供应商

veterinary ['vetrənəri]　adj. 兽医的

vigorous ['vɪɡərəs]　adj. 充满活力的，果断的，精力充沛的，强壮的，强健的

violation [ˌvaɪə'leɪʃn]　n. 违反，妨碍，侵害，违背，强奸

virtually ['vɜːrtʃuəli]　adv. 事实上，几乎，实质上

viscosity [vɪ'skɑːsəti]　n. 黏性；黏度

viscosity decreasing agent　n. 降黏剂

viscosity increasing agent　n. 增稠剂

visible light　n. 可见光

visual ['vɪʒuəl]　adj. 视觉的，视力的，栩栩如生的

Visual Artist　视觉艺术家

volume ['vɑːljuːm]　n. 量，体积，卷，音量，大量，册　adj. 大量的　vi. 成团卷起　vt. 把……收集成卷

voluntarily ['vɒləntrəli]　adv. 自动地，以自由意志

Voluntary Cosmetic Registration Program　化妆品自愿注册计划

W

warrior ['wɔːriər]　n. 战士，勇士，鼓吹战争的人

water hardness　n. 水的硬度

wavelength ['weɪvleŋθ]　n. 波长

wax [wæks]　n. 蜡，蜡状物　vt. 上蜡　vi. 月亮渐满，增大　adj. 蜡制的

wealth [welθ]　n. 财富，大量，富有

website ['websaɪt]　n. 网站

well-being ['welbiːŋ]　n. 幸福，康乐

whale oil　n. 鲸油，鱼油，鲸脂油

wholesale ['həʊlseɪl]　n. 批发　adj. 批发的　adv. 大规模地，以批发方式　v. 批发

wise [waɪz]　adj. 明智的，聪明的，博学的

woody　木质的

word-of-mouth　adj. 口头的，口述的

workhorse ['wɜːkhɔːs]　n. 做重活的人，驮马，重负荷机器　adj. 工作重的，吃苦耐劳的

wrinkle ['rɪŋkl]　n. 皱纹

Y

yeast [ji:st]　*n*. 酵母，泡沫，酵母片，引起骚动因素

Z

IP Code　邮政编码

附 录

 附录 英语构词法

 附录 化学元素和化合物的英文命名

 附录 化妆品颜色中英文对照

参 考 文 献

[1] 付明明. 美容实用英语. 上海：复旦大学出版社, 2019.
[2] 贾长英. 化工专业英语. 北京：中国石化出版社, 2018.
[3] 张卫华, 余芊芊. 实用美容英语会话. 武汉：华中科技大学出版社, 2017.
[4] Kazutami Sakamoto, et al. Cosmetic science and technology. [S. I.]：Elsevier, 2017.
[5] 张艳红. 美容美发实用英语. 北京：中国人民大学出版社, 2015.
[6] Leslie Baumann. Cosmeceuticals and Cosmetic Ingredients. [S. I.]：McGraw-Hill Education/Medical, 2014.
[7] Gabriella Baki, Kenneth S. Alexander. Introduction to Cosmetic Formulation and Technology. [S. I.]：Wiley, 2015.
[8] 高瑞英. 化妆品专业英语. 北京：化学工业出版社, 2013.
[9] 江涛. 听懂英语才会说：美容健身篇. 北京：石油工业出版社, 2010.
[10] 吴红. 精细化学品专业英语. 北京：化学工业出版社, 2008.
[11] 浩翰, 钟乐平. 迷上说服饰美容美发英语900句. 北京：中国宇航出版社, 2006.
[12] 范广丽. 美容美发英语. 北京：中国劳动社会保障局出版社, 2005.
[13] https://hkmb.hktdc.com
[14] https://www.marketingtochina.com
[15] http://www.historyofcosmetics.net
[16] https://cosmeticsbusiness.com
[17] https://www.loreal.com
[18] https://www.fatbit.com
[19] https://www.cosmeticsinfo.org
[20] https://www.fda.gov
[21] https://ec.europa.eu
[22] https://colipa.eu
[23] https://www.cirs-reach.com
[24] https://www.cir-safety.org
[25] https://www.dermatologytimes.com
[26] https://www.femaflavor.org
[27] https://www.micamoma.com
[28] https://www.webmd.com
[29] Zoe Diana Draelos. Cosmetic efficacy testing. Dermatology Times, 2018, 39 (12).
[30] https://www.nmpa.gov.cn
[31] https://kns.cnki.net
[32] https://www.medicalnewstoday.com